Do-It-Yourselfer's Guide to Home Planning & Construction

No. 888
$7.95

Do–It–Yourselfer's Guide to Home Planning & Construction

By William Clarneau

 TAB BOOKS
Blue Ridge Summit, Pa. 17214

FIRST EDITION

FIRST PRINTING—AUGUST 1976

Copyright © 1976 by TAB BOOKS

Printed in the United States
of America

Hardbound Edition: International Standard Book No. 0-8306-6888-8

Paperbound Edition: International Standard Book No. 0-8306-5888-2

Library of Congress Card Number: 76-24781

Foreword

I'm not a carpenter, an architect, an engineer, or a building contractor. Aside from a single rickety treehouse I nailed together as a child, I'm an alien to the construction industry. But recently I built a house.

This house isn't a playhouse, a one-room cabin, or a semifunctional artwork. It's a moderately complicated building, even by professional standards; within its rustic cottage atmosphere is a full-size three-bedroom house, sitting atop a specially engineered hillside stilt foundation on a steep, heavily wooded slope.

I did all the work in eleven months with almost no assistance. I acted as the carpenter, the plumber, the electrician, the furnace installer, and the supervisor. And being inside the boundary of a major U. S. city, all my work was carefully inspected and okayed by building inspectors in strict accordance with standard building codes.

So although I'm not by any means an expert on housebuilding in general, I know very well what it's like to build a *first house* from scratch, without prior experience in housebuilding or any field related to it—something very few professionals can recall.

I wrote this book as one who built a house not primarily as a construction project, but as an exploration of his personal limits, making sure at each point where the new knowledge came from, noting in the best experimental spirit the forces acting on the process.

To test *myself* meant involving others only in controlled ways. So, for instance, when I might have taken a problem to a relative who designs houses and asked him to figure it out and tell me what to do, I purposely avoided his aid and thrashed through the difficulty on my own with books and other open-to-all sources of information—workmen, salespeople, bureaucrats—making decisions on the data available to me as a nonprofessional, seeing where the information came from, and gaining an awareness of exactly which bits of knowledge were necessary.

It turned out that two major kinds of information were exceptionally hard to find, both of them quite general. Both were patterns that emerged only late in the project, from a large accumulation of experience, and whose lack made everything more difficult, emotionally and physically.

One is an awareness of the invisible parts of a house—the million-plus vital and time-consuming details a contractor normally handles, as well as those added for a self-built house, where the time spent learning, dealing with people, searching for materials and information, worrying, and laughing at yourself can be much greater than that spent on craftwork itself. Blind to such considerations at first, I expected the house to take only a month or two to complete. Each day would be spent sawing and hammering, while the materials, permits, instructions, and good weather would appear as needed—*by magic.*

After city hall conflicts, scrounging for used materials, worrying about the weather, studying building books, driving thousands of miles on errands, and diffusing my energies in hundreds of other ways, the picture became painfully clear.

The second hard-to-acquire knowledge for the novice builder is a technical picture of the house itself. A *really* inexperienced person in construction (as I was) has no idea of the relation of a house to its parts. Even after reading several books about wood-frame houses, it is difficult to connect the concept of the house-one-lives-in with the house-in-pieces-from-the-textbook well enough to feel confident taking action with a saw.

A first-time builder needs to integrate the two perspectives to be able to oversee the details of construction and to have the requisite faith that each step of action will actually lead to a finished house. At each new phase of construction, a major obstacle was getting myself to believe I wasn't the victim of my imagination, since the similarity between the boards at hand and the appropriate portion of a finished house was very slight.

This book is written to provide overviews of both the psychic experience of housebuilding and the physical assemblage known as a *house*. The early part of the book focuses on finding a location and dealing with building inspectors, rising material costs, interesting characters, and significant events, problems, and rewards.

The remainder of the book concentrates on the technical aspects of housebuilding, describing from a naive layperson's perspective the forms and functions of the basic house parts —walls, floors, roof, ceiling, and mechanical systems.

This includes a collection of miscellaneous facts, tricks of the trade, notes on house kits, and such information that is likely to be useful.

William Clarneau

Contents

1 **Why I Did It**   1
Advantages—Experiences—Acquiring Information

2 **Plans & Designs**   1
Helpful Friends—Plan Books—Prefabricated Houses

3 **A Place to Build**   3
Related Matters—Land Costs

4 **Paperwork**   4
Financing—Back to the Bank—Permits and Licenses—Back to
City Hall

5 **Buying Materials**   5
Comparing Bids—Buying Lumber—New Or Used?—Digging and
More Materials

6 **Building the Foundation**   6
Finding Tools—The Foundation—Concrete—Concrete Versus
Cement—Contractors

7 **Carpentry & Framing**   8
Carpentry—Framing—Framing Methods—Framing Tricks

8 Roofing   **125**
Designs—Exposed Beams—Variations

9 Enclosing the House   **147**
Fireplaces—Windows and Doors—Doors—Walls—Ceilings

10 Plumbing   **169**
Plumbing Diagrams—Plumbing Systems—Waste Systems

11 Electricity   **193**
Materials—Fundamentals of Electricity—Wiring—Wiring Hints

12 Heating the Home   **209**
Heating Systems—Insulation—Conserving Energy

13 Moving In   **219**
Finishing—Odds and Ends—Conclusion

Index   **231**

Why I Did It

From the time I was a small child, watching workmen build the houses in my tract-house neighborhood, I thought someday, when I grew up, I too would build a house. The idea lay dormant in the back of my mind for many years. But finally, I made some free time for myself and saved money; I had the chance to turn my dreams into actuality, and I took it.

Housebuilding fulfilled some other needs for me as well. Life was getting too secure and planned; I was ready for a genuine, new challenge to attack. I wanted to do something I hadn't done before, something I wasn't sure I could do, to take a risk. I wanted to test my notions of, and pit myself against, the business world, government, indifferent Mother Nature, and my own inadequacies. I wanted to experience several different occupations, especially that of independent businessman, to set my own working hours, and above all to take more of the power controlling my own life and explore the unknown ranges of my ability.

I had to follow the example of my pioneer ancestors and strike out on my own by adult standards. By finishing the house with little help or professional experience, I could proclaim my adulthood, convince myself that I could measure up to any man, take care of myself in the real world, and gamble on myself with confidence. Those workmen of my memory would be my equals.

With the spirit of construction in my blood, I wasted no time wondering why not; I thought of several reasons why I would succeed. I remembered several parents of friends who'd built the houses that my friends grew up in. One of my close relatives had even built a house once. If these people could do it, so could I.

I ran into a friend who'd recently been building houses, and he claimed it was quite easy to learn. I even met someone who offered his assistance, though he didn't come through with it. I'd heard of houses being framed in six days—it sounded very simple. I considered myself a self-reliant, independent person, confident that I was equal to the task. I'd read some how-to books that made things sound quite simple; I could complete the house, I reasoned, during the summer, while the weather was nice and working outdoors would be pleasant, and have an all around good experience.

ADVANTAGES

Aside from my own motivations, I saw that building my own house had many practical advantages. Financially, a self-built house can be an excellent investment. It can make money through fantastic savings on mortgage interest, protect against inflation because housing values rise much faster than inflating prices, reduce income taxes because homeowners can deduct interest payments from their income, produce a house at about half the new-house cost since the labor is donated free, and even put an end to rent payments forever—if you can buy the materials without borrowing money.

You can live where you choose in a house you select or design for yourself. And the experience of grinding through a million details is an unbeatable teacher of the value of persistence.

I also saw housebuilding as a unique opportunity. Housing is one of the few fields where technology hasn't closed the door to small beginners. A single person or a small group would have enormous trouble trying to construct a modern automobile from raw materials. Thousands of dollars worth of machine tools and skill in their use would be an absolute necessity; designing would take a staff of special engineers; assembly without automated methods would drag on forever. To simply purchase the ready-made parts and assemble them would cost many times the price of a new model from the factory.

Industry has generally taken advantage of the economies of mass production, and thereby made products available that no one could otherwise afford.

House construction, still in the technical dark ages by comparison, is open to the individual. Of course, the fact that mass production hasn't yet replaced building tradesmen is a monument to their efficiency. But unlike the auto, electronics, or aircraft industries, the technology in building houses is very basic; there are many times fewer parts, and the tools are common, inexpensive, fewer, and easy to use. A house can easily be made with less than $100 worth of tools. The careful beginner can match the work of most experts; it just takes him or her a little longer.

Also, the supply chain in housing allows for independent work more readily than does the auto business; a lone housebuilder can get basic material at the same rate a contractor pays—but no individual can buy a carload of raw steel at General Motors' price.

This ease of entry for the individual only lasts while housing stays outside the realm of mass technology. Advancing production methods, while lowering costs, will eventually make it cheaper to buy than build a house. But until then the present system leaves an alternative for anyone willing to invest some time and effort.

EXPERIENCES

I've experienced city hall redtape, building and zoning codes, and inspectors. I've dealt with wholesale distributors and retail stores, fought with the weather and worried over the local ecology, scrounged through used material markets, shopped for subcontractors, listened to a hundred real estate salespeople and took their tours, hassled with banks, considered becoming a licensed contractor, studied building books for days on end, had sleepless nights, hammered 10,000 nails, inhaled sawdust, sweated over plumbing pipes, bolted bolts, built retaining walls, driven 5000 miles on both fruitful and fruitless errands, and played the parts of businessman, skilled tradesman, engineer, architect, truckdriver, ditch digger, weightlifter, student, lumber carrier, door and window installer, and sheet-metal worker. And all the time not as a professional, but as an average person without any preconceived notions, totally new to construction.

I was even seriously offered jobs at different times as an apprentice sheet-metal worker, a concrete worker, a carpenter, a salesman, a building wrecker, and a general contractor.

But at the outset I was exceptionally naive.

My wife relates the story of her younger brother who as a young child was fascinated by balloons. One day he was asked if he knew how to make one. After thinking for a moment, he made a circle gesture with his hands and said, "You take some color and make it around." Just so, I used to think that to make a house you took some wood and "made it around"—to form four walls and a roof. But compared to my imaginings, the complications were immense. In fact, my ignorance itself produced two of the most basic problems I encountered. I wrote this book to help save others from these fates born out of inexperience.

ACQUIRING INFORMATION

As I first began to think seriously of a self-made house, I hunted down some information on the subject. I found the how-to books overflowing with details on how to make a sill plate, a header joist, rafter, or whatever. Houses were broken into tiny, digestible pieces and strewn through several texts. Looking at this great array of house parts scattered here and there, I got an anxious feeling—as if I'd taken a watch apart and couldn't see how the pieces went back together.

On the one hand, I had the image of a whole house in mind—a regular old *HOUSE*, a comfortable, warm, dry place to live; the thing you get by taking wood and "making it around." On the other hand, I pictured this dismantled version—an erector set or unmade jigsaw puzzle that the books used for illustration. I couldn't find a text that put the two together, that showed the completed jigsaw or the watch parts back in place. So there I was—dependent on the piece-by-piece books for a guide to housebuilding.

The how-to books assumed the reader would faithfully and unquestioningly follow all the details they set forth. But trusting several thousand dollars, all my working energy, and my future home to someone else's explicit directions made me quite uneasy. I wanted to know where I was going and how each step I performed related to the overall product.

Had I built a house before, I could have formed a mental image of the assembled jigsaw with the parts exposed. But as it was, I felt like I was stumbling blindly forward. Perhaps the books were unreliable; or, if they weren't, perhaps my house was somehow different in a way that made the guide inapplicable.

I read lots of books on housebuilding, plumbing, electricity, and carpentry, to name a few. I studied pictures from house magazines, went to houses being built to watch the tradesmen working there and ask them questions, took careful note of finished structures for clues to their construction, and asked salesmen to share their knowledge of their products. I exhausted every possible source of information to find out what I needed. But I was always doing something new, and each successive part I worked on was a fearsome challenge, an unknown territory where I couldn't be certain that what I did was right; no matter what the how-to books or nearby workmen said, I had no personal experience to trust in. Without my own basic understanding of the whole assembly, I had only my blind faith in other people and their piecemeal writings to rely on.

Ironically, I got a feeling for the whole thing, in relation to its parts and processes, only when the house was practically done. Then the new awareness seemed so trivial—as if I'd never been without it—that I could see why it was easily overlooked in housebuilding texts. Now the awful, anxious uncertainty of doing each thing for the first time vanished; at last I had a general understanding of the whole, assembled jigsaw house—a mental image of the watch with all its pieces back in place and working. The rules and processes I followed blindly became the simple outgrowths of my own common sense.

Part of this book helps to set the irony straight—to help other first-time housebuilders get a feel for the whole thing early in the process, long before they finish their first house. It presents the house from the beginner's standpoint—the naive view that sees the house as nothing more than four walls and a roof, a *wood balloon*. It starts with what you most likely know already, the basic house concepts, and explains what parts make up the thing you see. Most how-to books tend to ignore the fact that you already have an image in your mind of what a house is. They begin with parts of houses and require you to memorize their assembly procedure, leaving up to you the understanding of a house as you already know it in relation to their jigsaw puzzle.

Unlike them, I use your present knowledge, starting with the parts you know—walls, floor, ceiling, roof, bathtub, lights—and tell you how they work, why they are there, and what they are made of. Once you know the basics, the details will fall into place by sheer common sense.

The other major setback caused by my own inexperience has to do with the enormous undertaking do-it-yourself housebuilding represents. I'd blown up plenty of balloons and built model airplanes, each usually in a short while. But I'd never built a house; I didn't realize how long this chain of activity would take—a month or two at most was my original plan. Instead, it commandeered my life for almost a year of constant toil. Even if I'd had a contractor dictate a long detailed list of instructions of what to attach to what from start to finish, I couldn't have avoided one mistake—I couldn't know how long each step would take for *me*, an inexperienced worker.

I couldn't know the kinds and quantity of events I was going to face by building my own house while learning in the process. No supply of technical books read, people questioned, or sample houses visited could indicate what my life would be like during this undertaking.

Part of this book is, therefore, designed to illustrate the experience of single-person, first-time house construction. It is a narrative combining some of my most illuminating experiences, the people I ran into, and the unexpected problems. Although not encyclopedic, it's meant to give a strong impression of the kinds of happenings that make up housebuilding, and the life a solitary housebuilder might expect to lead—topics not relevant to ordinary how-to texts.

So, actually, this book doesn't tell you how to build a house. Its purpose is to clarify what houses are so you can guide yourself through house work. It's designed to take you from the wood-balloon stage of awareness (or nonawareness) to a general understanding of houses, especially self-made ones. Once you're there, your pool of "common sense" can guide you, as you pluck details from how-to books and other information sources you need.

Part of this book contains some things I found quite valuable when I was working on the house—some tricks, hints, and general notes on houses and housebuilding that don't fit in anywhere else, as well as a few notes on *building for the energy crisis*.

Plans & Designs 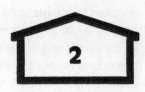 2

Designing a house would be relatively simple for me now. But when I first thought of building a house, I had little idea of what was involved. I couldn't have designed a doghouse. I thought architects might help, but discovered them to be a resource I couldn't afford—charging 10–12% of the finished-house cost as their fee. Several magazines and many independent services sell plans already drawn, however; I decided to take advantage of their low price—$30 to $100—and started looking for a suitable design.

My wife Diane and I spent days and days exploring books, old magazines, and plan catalogs for something we liked and could afford. We were awfully picky, though, and couldn't find a thing. I wanted an absolutely perfect house, one with everything in just the right place. After all, I was only going to build one house in my lifetime, so it might as well be as perfect as possible. But it wasn't easy. One house would have a good floor plan, but the exterior siding was atrocious, or the roofline was ugly. Another would be appealing inside and out, but too large. Still another would have a cozy fireplace, but a ceiling we didn't want, or too little closet space, or a large, wasteful garage, or no bathtub. Sometimes we would find a house that seemed perfect in the photograph, but wasn't available as a standard plan. I didn't know the first thing about what holds a house together; and I couldn't visualize the ways to modify any plan to suit our taste. So we continued to look for the perfect plan.

After scouring through a thousand stock plans to no avail, I hit upon the idea of beginning with a stock plan, making changes to produce the perfect house for us, and taking the whole thing to a professional builder to make sure it would work correctly. We set about collecting all the features we wanted in our home. I measured rooms in every house we visited to see what sizes our floor plan should use. I got so addicted to measuring and comparing, it became a joke. Our friends would greet us as we arrived, and I'd immediately be on my hands and knees in a bathroom, bedroom, living room, or kitchen, measuring the floors like a bloodhound sniffing for a clue. We measured kitchen counters, fireplaces, window openings, ceilings, doorways, halls, and everything imaginable.

Compiling all these facts on paper was a far cry from the design of the house, though. I was filled with doubts. I didn't know a thing about building codes, plumbing, or the structure of a house itself. Above all, I was ignorant of what any changes in the plans would cost. Indeed, I had only a general idea that self-built houses can save money, but I didn't know what anything would actually cost.

We went on collecting ideas. I dragged my feet, hesitantly looking for some land, waiting for my anxious mood to change. The thought of giving up the whole thing started to sound like a welcome relief.

Then I met someone who lifted me, by his example, from the quagmire of inaction I found myself sinking into. Passing a house under construction very similar to what we wanted, a relatively small house in a wooded area with a cottage-like exterior, I stopped to investigate, calling into the house.

"Hello," I said. A single hammer-tapping stopped. "Hello?" I repeated. A young man, not much older than myself, peeped his head out the door. He wore blue jeans, a flannel shirt, and a carpenter's tool belt. I thought this workman might know something about plans.

"Hi," he said.

"Mind if I come down?"

"Sure," he said, waving his hand to invite me in. I slid down the slippery clay hillside, and we introduced ourselves. Inside, I looked around at the bare skeletons of walls, a work table with a power saw sitting on it, some lumber scattered around the floor, small piles of sawdust, and power cord draped through a window opening, leading outside.

"Looks like you're the only workman here today," I said.

"I'm the only workman here every day. It's my house," he said.

Incredible, I thought. Could he actually be doing what I had been agonizing over?

"You mean you're doing everything yourself?" I asked.

"Oh, sure. Not much to it, really. Just a lot of grindy stuff."

"Are you a contractor or carpenter by profession?" I asked, thinking that surely a house this big, with all these professional tools around, would take experienced hands to build.

"Oh, naw, just puttering around. I'm a fisherman in season."

"Well, then, you've built something before this, haven't you? This isn't your first building project, is it?"

"I built a house last year, just like this one, only a little smaller—but that was my first one. It was a lot more fun. Doing everything for the first time, you're amazed at what you can do. It's a real kick." He fumbled through his wallet, retrieving a photograph. "Here's a picture." The photo was of a perfectly built little cottage—exactly what we were looking for. Maybe he would help us with our plan, or sell us his, I thought.

"I really like your plan," I said. "I've been trying to find one like this myself. I like the high, steep cedar-shingled roof, and the rustic siding. I'm thinking of building a house by myself, too. Did you have someone draw the plan for you...an architect?"

"Oh, no," he said modestly, "I did it myself."

"Didn't the building people give you any trouble? I thought building houses within the city limits was supposed to be a lot of trouble."

"Well, it's my own house, y'know. A guy can do pretty much anything he wants with his own house. They made me get an engineer to certify the plans, but they took it okay. Looked at it pretty hard, too. They called it an '*atypical*' design, with all these bolts and things," he said, pointing to some bolts where the walls and ceiling came together. "But they gave me the permit, and that's all I cared about." He patted a nearby beam. "It's solid."

He seemed so casual about the whole thing, I couldn't believe it. I noticed some pipes and wires snaking through the skeleton walls.

"And you're doing your own wiring and plumbing too?" I asked.

"Yup, nothing to that either. It's all common sense."

Here I'd been worrying about a thousand tiny details, and he just went ahead and did it; I was ashamed of my timidity, but inspired as well. He knew what he was doing, and had a plan that the city building inspector had already passed; his help could be invaluable. We talked for a while and he agreed to sell me his plan, as well as some of his experienced assistance (when I got stuck), for $200.

My anxiety dissolved. I found help, a perfect house plan with cozy loft bedrooms and a fireplace, and inspiration in his independent attitude. He even told me the house should cost no more than 25% of what the finished value would be. I scrambled back to the car and hurried home to tell Diane the good news.

HELPFUL FRIENDS

While searching for land, I paid a few visits to my new builder friend. His motto seemed to be: "Well, a guy can do pretty much anything he wants." I'd ask him things like were there any kind of code requirements about how many windows a house must have, and he'd say, "Well, a guy can do pretty much anything he wants." I'd ask whether he had calculated the amount of wind the house could withstand, and he'd respond the same way. I understood his point that the inspector isn't usually hard on a person building his own house, but I suspected his attitude might stand in the way of safety or, at best, was overly dependent on the inspector's good nature.

One day he casually mentioned that he had decided to add more wood to a certain roof beam, since the weight of the roof was making the walls lean outward. The fact that he had designed this thing himself, and that he had pushed it through the building inspector's checking process, flashed through my mind. My faith in him fell several degrees.

Yet his easy-going attitude set the pace for me. He made everything sound so easy, it inspired me; all was just "common sense" and, if problems arose, "a guy could do anything he wants." I compared him with a much loved and respected uncle of mine who always managed to solve any problem he faced simply by assuming (as I watched him do many times) that the means for its solution always existed. Then he would create his solutions forthrightly, without getting upset over the obstacles that were in his path. The sort of person who worries over doing everything perfectly with ease will meet continual frustration and slow progress when building a house. And if I hadn't developed a way of knowing what was adequate, moaning over imperfections instead, the house would be just now beginning.

By the time we bought the land, the notion that I was planning to build a house had circulated through my family and circle of friends. Frequently, I would hear stories about other people building their own houses—or friends of friends who were engineers, contractors or carpenters who had offered help through the "grapevine," if I needed it. It was amazing how this network of interested people seemed to spring up. Just as a newly learned word starts popping up everywhere, it seemed like half the people I knew were suddenly associated with housebuilding. I contacted one such person, a friend of one of my parent's coworker's daughter-in-law. He had recently graduated from a course in construction engineering and had designed his own house. I showed him the plan I bought and asked how the wall problem might be corrected. By the time he explained it, I decided my builder friend was a less-than-reliable source of help, perhaps even a hindrance. It turned out that the building lot I purchased was too narrow to build the house on anyway, without some drastic changes. So I resolved to buy a standard plan, after all, from a bona fide plan service. Making the necessary changes, I would come up with a reliably designed house that would fit properly in the available space.

My close friends were all excited, building up a psychic momentum for the project. My family thought I was joking when they heard my plans, or if not joking, at least mildly crazy. Those who had experience in the building industry were curiously tight-lipped. When I mentioned in June that we planned to be finished by September, someone said, "This September, or a year from this September?" I thought it was a quip (though it turned out to be very close to fact) and laughed at what I saw as a typical lack of confidence in youth.

Another friend, a contractor for many years, visited the land just before we started. He was pessimistic too. He vowed he would never contract to build the house we intended on the site; I chose to hear his statement as a remark of admiration for my courage. I later learned that he had masked his feelings, and had honestly thought it couldn't be done by such an amateur. Another person simply shrugged when he saw the lot, and said, "Well, you've really got yourself a project."

Almost everyone was incredulous when I announced my plans to do everything myself—including the foundation. The uniformly negative response, especially about the foundation, acted to set up within me a large (and unwarranted, I found) fear of doing it.

Though I never took these admonitions too seriously, they were far from helpful in allaying my anxiety over this new and unknown area. When it came time to build the foundation, I didn't see any reason why I couldn't do it myself. My memory kept playing over and over the visions of those people, sighing in hopeless resignation when I said I was doing even the foundation myself. All I had in answer to these "voices of experience" was my notion that there wasn't any magic involved and, as long as I did things according to the book, there would be nothing to worry about.

But I worried anyway. Who was I to challenge such complacent warnings? I had only my sense of reason to rely on—my emotions weren't convinced. So I went ahead without much confidence.

24

PLAN BOOKS

While looking for a plan, I contacted another old family friend, the contractor who had built my parents' house. My wife and I visited him one night, with such questions on our list as: What do furnaces cost? What do sewers cost? Plumbing systems? Wiring? Is there a limit as to how far a person can build a house from the road? What kinds of requirements must a builder worry about? Can I use a wooden foundation? Are there any standard room sizes?

Soon after we arrived, it came up that we needed a plan. He produced some plan books—those, in fact, that he used in building my parents' house 20 years before—and we immediately recognized them as the forerunners of the present day plan-service books, with all the same basic houses and slightly older looking siding. He pointed out the design for my childhood home. Referring to the other houses on that page, he indicated how a few boards added here or there gave a different appearance and a new architectural effect to the basic house. All the houses in my old neighborhood were really identical beneath the various facades! He proceeded, my childhood illusions of neighborhood architectural variations now shattered, to recommend a small house for me to build, suggesting different ways it might be altered to give a "unique" effect.

Unwilling to let our naive house fantasies bow to his experienced advice, we quietly ignored his remarks. The exterior walls of the house he proposed were a simple rectangle, with no extra corners to waste materials and time; its roof was shallow-pitched for easy working conditions; and the house sat on level ground where the foundation would be simple. Overall, the house was small and easy to build.

We mentioned that we were thinking of high, exposed-beam ceilings, wood paneling, and a wooded hillside location; but he casually brushed these ideas aside. He warned us not to pay more than 35¢ per square foot for wallboard; said that of course we wouldn't want to get mixed up with a high ceiling, because of the heating costs; and dismissed the hillside location and paneling altogether, as if such silly notions were a quirk of madness best ignored. He was very nice, but our total ignorance of the subject made it difficult to communicate our thoughts to him, as well as to get a feeling for the practicality of our imaginings. It turned out that had we followed his advice, the building time would have been cut in half.

At any rate, we left our old friend's house slightly wiser, a bit put off by his unheeding devotion to tract housing without understanding why, and enlightened as to why the plan services were so devoted to tract plans—all the professional builders used them.

After searching through more plan catalogs, we finally found a plan we could use and bought it—for $40. The plan service also sold a materials list—$5 more—as well as a "typical plumbing and wiring diagram"—another $10. I planned to take the materials list to a lumberyard and order everything at one time. I thought the plumbing and wiring diagrams would show how the plumbing and wiring fit into this particular house. It turned out that they were only general descriptions of how the plumbing goes into any house—not a detailed instruction sheet for the house I was going to build.

Plan services have, basically, two kinds of plans. Most are for 1950s type suburban tract houses; a few are what they term "vacation" houses, with A-frames and other less orthodox designs, using more wood and windows than the tract houses. We had a special aversion to tract-type houses, however, and blanked them from our mind at once.

According to the plan-service catalog, our selection was a vacation house. It had a few drawbacks for year-round occupancy in our setting, however. It was designed for the high mountains, including enormous, expensive roof beams meant to hold up ten feet of piled-on snow. The floor plan had a small upper level with a loft bedroom, and a small bathroom downstairs. It was meant for level ground, and the windows faced sideways, away from our view.

The plan needed several modifications, including a special foundation for the hillside. I thought I'd simply add a few boards here and there, shift the bathroom upstairs, put the windows on the appropriate side, and give the bundle to a structural engineer who could okay it and draw in the hillside foundation. But, alas, the engineer had to redraw the whole thing, since the law in my state says an engineer cannot certify anything he isn't the sole author of—he couldn't just draw the foundation and put his seal on it, he had to draw the whole house. I hadn't the faintest idea of how to draw a suitable foundation myself, so I was at his mercy. I gave him a detailed drawing to work from, though, so he only drew in the rough necessities as he redesigned the house—but these services cost more than $200, a painful addition to the $55 I paid to the plan service, thinking then that the design costs were covered.

The engineer didn't quite know what to make of me. He said he normally worked directly with architects or contractors who knew the construction lingo. But I hid my ignorance well; he agreed to help me. He tore himself away from one of the skyscrapers he was designing, and whipped out my plan for what he figured was a reasonable fee—$16 an hour.

He said that he had to have another engineer's soil test to tell him what kind of foundation to use. Luckily, I didn't have to find another engineer. I obtained a copy of the soil report from the owner of the property next to mine, who had already paid a soil engineer for a report on his land.

PREFABRICATED HOUSES

Prefabricated houses are increasingly popular, judging from the large number of companies making house kits. For most prefabricated houses, however, the cost for all the pieces and assembly instructions is as great as that of a ready-made house. And you still have to put it together! Small houses—less than 900 square feet—are especially expensive if you buy a kit; you pay dearly for the instructions, perhaps 3—4 times what the raw materials are worth.

Most kits include only the rough carpentry, leaving the foundation, plumbing, wiring, insulation, roofing, interior wall coverings, cabinetry, floor coverings, gutters, driveway, weather-stripping, and paint for you to do. They take advantage of the illusion most beginners have that four walls and a roof is a house, when really it's less than half the total cost of the house, and the easiest half at that, since none of it shows in the finished house.

You can roughly compare any house plan that you are interested in to a similar prefabricated kit. The cost for land, sewer, water, building permits, and driveway access is the same for any house on a given piece of land. The house itself, the foundation, excavation, lumber, windows, doors, roofing, mechanical systems, finishing materials, and miscellaneous materials are the parts to compare.

Take your plan to a lumberyard and find out what the framing lumber, roofing, and building paper will cost, and compare that to the cost of the prefabricated kit. If the kit doesn't include roofing, for instance, subtract that item. Normally, the kit only includes framing lumber, so the comparison needn't go much further. Most kit companies get good bulk-rate prices on their lumber, so it might not cost much more for the kit than the lumber—but you must compare.

The plumbing, wiring, heating, and other things the average kit leaves out can be priced easily. Take the plan to a contractor for each system of the house and cut the bid you get in half, since labor is always at least half the cost. Half of the contractor's price will probably be near the materials cost, unless it's a ridiculously high estimate.

Depending on your interest, it might pay to learn enough about parts of the house, such as the plumbing, for instance, so you can tell what parts you need to compare should a kit contain other parts.

The major advantage in buying a prefabricated kit is having all the parts when you start out, eliminating most of the time you would otherwise spend buying them. Carefully researching all the house parts, the processes by which they go together, and the best prices available will give you your own kit.

With all the parts ready beforehand, a prepared person should easily be able to build a 400—500 square foot "starter" house in two months, with the following assumptions:

1—The land is cleared and level.
2—The foundation consists entirely of separate footings (with posts) needing no forms and resting 2 feet deep in the ground.

3—Floor joists are 4- by 8-inch beams on 4-foot centers, with thick plywood subflooring.

4—Windows are prebuilt and doors are prehung.

5—Wall frames are 2- by 4-inch studs with 16-inch centers.

6—Prebuilt roof trusses are used for a shallow roof pitch.

7—A metal fireplace or Franklin stove is included.

8—Forced-air furnace is included with flues and ductwork.

9—Copper tubing plumbing is used, with cast iron or plastic waste pipe.

10—Lumex flexible wiring is used.

11—Blanket wall insulation is used.

12—All permits, materials, and tools are obtained beforehand.

13—The sewer and water lines are each less than 150 feet long.

14—Sewer pipes are cast iron no-hub, and water pipes are galvanized steel or copper.

15—Interior wall coverings are 1- by 6-inch wood panels (plasterboard applied by a contractor would save a week's time).

16—Ceilings are exposed roof sheathing and plywood.

17—One hour per eight-hour workday is allowed for errands.

Except for stringent codes, laws, and regulations imposed in all populated areas of this country, requiring the builder to stay within the confines of accepted building practices, it would be possible to build houses from mud and twigs, or other cheap materials, and have a two-bedroom house for $500. Ken Kern in *Owner Built Home* goes into more detail on building such inexpensive, unique, efficient houses by yourself, using unconventional materials and methods. Making use of these economies is impossible within the confines of the normal building codes, however.

Building a house of even the most conventional sort is a mammoth challenge. Without worrying about innovation, wood-frame house construction is probably the simplest avenue to follow, with everything standardized and all the local salespeople and tradespeople around to help when you get stuck.

A Place to Build

Finding land to build on is similar to shopping for a good used car, but much harder. Cars, at least, are small and portable; they can be bunched together in shopping areas, used car lots, etc. Vacant land is scattered everywhere, and photos seldom substitute for a personal inspection.

THE RIGHT LAND

The quest for land took quite a while, especially since we adopted the approach that, having a choice in the matter, only perfection would do—a forrested hillside with a view, within our budget. It was like trying to find that mythical "good used car," although good land neither arranges itself along an auto row nor is advertised on the late movie. Much land isn't advertised at all.

The land hunt led to architects who were holding the land for their own dream house, to people holding it as an investment, to some willing to sell but for a high price, to development corporations gobbling up the countryside, to farmers getting rich selling their farms in small pieces, to public foundations holding land in trust, to swindlers selling land illegal to build on, to people making 400% profit on land they'd held a year, and to one real estate agent who was so confused that he got us to the sign-the-papers stage before discovering, to our dismay, the owner had already sold it.

Checking out prospective lots brought me to city hall, where I spoke to city planners about zoning maps, and to public works engineers regarding water supplies, sewers, and building regulations. The land within a 20-mile radius fell into three counties, necessitating trips to three city halls for these details.

Each lead involved a phone call to the person in charge, followed by a trip to the land if it sounded suitable, a very time-consuming process. At first we thought that only advertised land was actually for sale; the rest was publicly owned, parkland, or their ilk. But no, most land was private and for sale for the right price. As long as any serious prospect involved a trip to the site anyway, I started working backwards, finding good land and calling to see if it was for sale—and if so, for how much. Of course, the latter meant ferreting out the owners' names and addresses from tax records. But in spite of the drudgery, this method was the one that finally worked.

We saw land of all kinds, from grassy farmland in five-acre tracts to wooded hillsides and bare city lots, searching from March through July.

Land hunting had its own whole new vocabulary. Sewers, septic systems, easements, zones, restrictions, rights-of-way, utilities, section maps, soil types, tax base, and many other terms assumed a special new importance.

Having by then searched for three months, we began to wonder if we wouldn't be able to carry this thing through. But I was still convinced that sooner or later, if we looked long enough and hard enough, we would find a landowner generous enough to part with his or her land at *our* price, thereby assisting a "nice young couple" in the building of their home.

After several more days of searching through municipal property-tax records to find the owners of attractive vacant land, the miracle occurred. I called the owner of a piece of land we really liked, and asked if it was for sale; it was—and for half the going price! Just like in the movies, we succeeded only after hope was almost lost, and my conviction proved itself true. Rather than being generous, however, the owner had parted with the land so cheaply only because she had bought it for much less, and realized a healthy profit anyway.

I had once thought real estate prices were like those on items in a department store—everything has a fixed price, and that's what you pay. But after a while I discovered that real estate is a bargainer's domain. I first became aware of the state of things when a friend and I looked at some property together.

We had an appointment with a somewhat aggressive salesperson who kept addressing us as "boys"—"What do you think, boys? Nice little place, eh, boys?" My friend inquired as to the *asking price*. He didn't say "how much is this place?" He said, "how much do they want?"

"The asking price is twelve thousand," she replied. "They've been offered eight, and refused." Cupping her hand to her mouth, as if to ward off eavesdroppers, and hushing her voice, she said, "I think they'll take nine-five."

This was way over my budget, so we thanked her and left. I've since concluded that many sellers make the asking price as outrageous as they can, in hopes that someone will come along and be so enthralled with the property that they will pay the owner's top price. Lack of money, however, kept me from making a hasty purchase. After a while, I simply told the agents, after I heard the asking price, that I only had so much money—did they think the owner would go for that? (I was telling the truth, or I might not have sounded so sincere.)

The power of my poverty came to me one day when I inquired about some property with an asking price of $7500. I said I only had $3000, assuming the agent would reply, in a ridiculing, condescending way, "sorry and goodbye." Instead, she said she would present the offer to the owner. Several days later, she called and said that "three thousand was indeed too little, but the owner would let it go for six thousand." I was amazed; and though I still couldn't afford it, I began to understand the business world.

There were many times when I would have made a purchase, high price or not, if I had the funds on hand. But the way it turned out was as the saying goes—"the rich get richer paying only for bargains." My attitude used to be that something was worth the price if I really wanted it. And most salesmen seem to take the attitude that if you're willing to pay too much for something, why not let you do it.

Aside from buying the land, I didn't really begin to care whether or not I did everything cheaply for quite some time. At first, it seemed strange listening to my housebuilder friend tell me how to save a few dollars here and there. The house was costing thousands; how could a few more dollars matter? Why not simply get the best and pay the price? Building the house was hard enough, without worrying about scrimping all the time on small savings. Eventually, I abandoned this attitude in favor of a more efficient one—but only when the money was almost spent.

RELATED MATTERS

The first entry on the long list of ingredients that make a house is the land. Searching for land to build a house on (*building site*) is far different than looking for some bare land to farm or have a picnic on.

The house must have a water supply. If a public water main runs nearby, wonderful. Otherwise, a well must be dug, which now costs about $10 per *vertical* foot. Wells can be anywhere from 10 to 1000 feet deep, depending on several things. Usually, the depth of nearby wells will give you a good indication of how deep you must dig.

Sewer Systems

A homeowner must dispose of sewage somewhere; it can't just run into a hole in the ground, at least not in most populated areas. Unless a public sewer runs nearby, a septic-tank system must be installed. You will need a permit from the local government to either install a septic tank or connect to the sewer. A septic tank costs around $1000 to install, and the permit to do so is usually cheap. Connecting to the sewer line, on the other hand, may cost a bundle just for the permit. My permit cost $650; but much less for the actual hookup.

The *sewer pipe* collects the waste from many houses and deposits it at a sewage treatment plant. The treatment plant can only serve so many houses; and today plants in many communities are unable to take on additional units. Sewer-connection permits are often "frozen" to prevent additional houses from feeding into the sewer.

A *septic tank* holds the sewage for a while, breaks it down to a liquid in a special treatment system, then pours the liquid into the ground. If too many houses in a small area do this, the waste comes back out of the ground. And a septic-tank system won't work on hillsides or in very hard soil. In some areas only a house that has two full acres can use a septic tank, since it can take that much land, experts say, to take care of the waste.

If you buy land where they won't allow either sewer connections or septic tanks, you can't build a house and you're in trouble. The *only* way to know for sure is to call the local government officials and ask them if the land is approved for a sewer or septic tank.

Zoning

Zoning is another consideration when buying land. In any city the planning commission or political machine decides what kinds of buildings can be built in each part of town. Some places are designated for shopping centers but not houses; some places allow houses but not offices; some allow apartments but not houses; and some allow combinations of these. If you want to build a house, make sure that the land you buy is *zoned* for residential housing. It sometimes happens, as a demonstration of bureaucratic inconsistency, that one piece of land in a particular area will be a zoned differently from surrounding land in the same area. You must check the zone for the piece of land you're considering, not for just the general neighborhood.

When you look across the land, wherever it is, you may not see many fences or property lines. However, every bit of land in this country is located on a map, where boundary lines from one piece of property to another are shown. In a neighborhood of houses, each house has its own *lot*, or piece of land, which is shown on one of the maps in the government offices.

Imagine the bare land that a neighborhood of houses sits on before the houses and streets were there. Originally, that bare piece of land was mapped as a single chunk, one large piece of land. When the people who built the houses came, however, they bought the big piece and divided it into a lot of little pieces. Each piece was to be the lot for a single house.

This process of chopping a large piece of land into little sections is called *subdividing*, and the neighborhood is called a subdivision.

Most of the land in a metropolitan area has been subdivided, vacant or not. A vacant area of land may actually be a number of subdivided lots, with different people owning each one. If you want to buy a piece of ground, you will have to have someone show you where the property dividing lines are, or obtain a map of the area from the local government.

The local zoning board, of course, has some rules about subdividing—rules that change from time to time. Basically, the one you need to be aware of is the *minimum lot size*, for a given area, that a house may be built on.

In addition to a minimum area requirement, many zoning laws require such things as a minimum width (e.g. the lot must be at least 50 feet wide at any point). You don't need to know these rules yourself; you need only know the legal description of the land. The officials can tell you if it is legal to build on the lot.

Developer's Restrictions

People who make the subdivisions often add their own restrictions, called *building restrictions*, on the lots. These stand apart from what the government may require. After looking at a few lots, I developed a routine series of questions, which went something like:

- Is the lot buildable?
- Is a sewer or septic-tank system available?
- Is there potable water available?
- Is there power available?
- Are there any building restrictions?

One salesperson answered that there were 10 pages of restrictions. Some of them were·

- No house can be built with less than 2000 square feet on the first floor.
- Houses must be designed by a specified designer.
- All front yards must be lawn grass.

Some salesmen said they "didn't know of any restrictions." In all cases, any such restrictions must be printed on the deed to the property. You can inquire with the county recorder (who has a copy of the deed) as to whether or not there are restrictions.

Normally, you don't have to check any of these things yourself until you decide to make a purchase; then you want to know exactly what you're buying. Trust a salesman only until it's time to sign the contract, then make sure you're getting what is promised before you sign.

Building Codes

You will also have to follow the building codes throughout the house, if you're building in a populated area. The inspector comes around for his final inspection when the house is completed, and gives you a certificate of occupancy. Without that certificate, you break the law by living in your house.

One of the first code requirements is that your house be positioned on the property a certain way, with a minimum distance to the property lines.

Permits are something I never thought too much about when fantasizing about building a house, but you have to get one every time you turn around, it seems. And permits can be expensive. Here are the costs for my permits:

Permit	Cost
Building	$56
Plumbing	$60
Electricity	$45
Heating	$15
Sewer	$1650
Water	$1140

LAND COSTS

Esthetics and practicality are at odds in choosing land. Not knowing anything of what was or wasn't practical, I decided basically in terms of esthetics. I chose a densely wooded hillside with a magnificent view. I doubt that I could have found a less practical situation.

Costs for land depend on quite a few things. The primary concern is how much the house will cost to build in one place as opposed to another. If two pieces of land have the same price tag, but one has the sewer and water lines already installed, the undeveloped piece is really much more expensive in the long run, since you'll have to pay to install the sewer and water lines. If one lot has a road leading up to where the house would be while the other doesn't, the one with the road is cheaper by the cost of a road. A lot that costs $5000 but needs a $4000 *piling* foundation is much more expensive than a $7000 lot that can use a *normal* foundation.

Location

After the house is built, the land it's located on will determine other kinds of costs you'll have to pay. Land in one area may increase in value more rapidly than land in another. It might be wise to spend more on one piece of land in order to increase your equity when the house is sold.

Some areas charge a higher property tax than others. Buying cheap land with high taxes may cost you more over the years than buying an expensive lot with lower taxes. The county tax people are the ones to consult regarding property taxes.

Finding Land

Tax records are a valuable resource in another way. Since every piece of land is owned by someone, and everyone has to pay tax on property that he or she owns, the tax people have records of who owns what. If you see vacant land, but it doesn't have a FOR SALE sign posted, you can locate the owners by simply calling the tax office (or checking the records yourself, they're open to anyone) and asking the name and address of the people who own the property. Usually, you have to know the legal description of the land, which you can figure out from the subdivision maps.

Real estate advertisements always seem to list the market value price. Only by hunting for a bargain can you find one. I found my property by sifting through an endless number of property records in the area where I wanted to live. Such opportunities aren't likely to come through professional real estate channels, since the professionals could buy bargains themselves and resell at the market value.

One specific source of land that might appeal to a novice builder who isn't picky about trees and seclusion is other home builders. Large homebuilding companies often subdivide land, build a few houses, and put the unused lots up for sale. They make a profit just by installing the utilities and streets then selling the land. The cost is usually quite reasonable when you consider that the water, sewer, electricity, and sometimes even the driveways are already included. As long as everything is legal, much time can be saved in such places.

Financing Land

If and when you find some land you like, buying it may be a problem without cash. Banks often have a policy against lending money for raw land, even though they will finance the house *and* land after it is built. They say it's to stop people from buying land on speculation. You can sometimes get a loan from a credit union, if you belong to one. Sometimes, the person selling the land will let you pay for it on a *contract* basis, at so much a month until the land is paid for. Otherwise, you will have to pay cash. If you buy land on a contract from the seller, you may be able to persuade him or her to let you build the house as you pay off the contract, leaving your money free to use in building the house. The seller may not agree to this unless you get a contractor's license, since he or she can lose the land if you run up debts and can't pay.

Once you own the land, a bank is more likely to lend you money for the building materials. Usually, however, you can only borrow an amount equal to the value of the house already completed. You need to have enough cash to build the foundation first, for instance. As soon as it's completed, you can borrow against its value, use the money to complete the next stage, and so on.

Paperwork

Building a house isn't all hammers and nails. There is a considerable amount of redtape that must be untangled. First of all there is the problem of money. Banks aren't too happy about lending money to a novice housebuilder, especially when that builder is unemployed, which he most likely will be if he has the time to build his own house.

Another problem is dealing with the local governing body. Full-time builders and contractors know how to deal with the bureaucrats at city hall. Some must be wooed, others must be dealt with firmly. People who are in and out of city hall on almost a daily basis get to know the people who work there—their moods and idiosyncrasies—and can deal with the frustrations. Those who are new at the game must learn to take things as they come.

Since all banks and governing bodies have their own rules and codes to go by, it would be impossible for me to consider every possible form, certificate, permit, license, or whatever else you may have to complete before you can get on with the actual building.

What I have provided is a sample of what I had to face, in hopes that you may be somewhat prepared for the battles that might be awaiting you.

FINANCING

Although some people have an abiding distrust for banking institutions, I had never felt compelled to share such sentiments. I had a checking account, a savings account, and had once even tried to get a job in a bank. I walked into the bank I dealt with, purposely oblivious to the guards and their battery of TV monitors, and consulted the wall directory. Noting its length, I awakened to the fact that the 20 floors above me were filled with banking offices.

"What sort of loan do you want?" asked the guard at the door.

"Well, I don't know," I said, "I'm building a house."

"You want real estate loans, then," he said, pointing the way. I marched past the savings counter, with which I was familiar, over to the counter for real estate loans. I stood at the counter for a few moments. I'd dressed up for this occasion, wearing some conventional clean clothes. My hair was just collar length then, though ragged despite my efforts to make it appear shorter.

A gray-suited woman met me at the loan counter, examined me through her silver-rimmed spectacles, asked my business, and said, "You'll have to speak to Mr. Brown."

I watched the people working, rustling papers, talking to customers seated at their desks, and giving directions to passersby stopping at the counter. The deep pile carpet and the air-conditioned climate removed all unpleasant airs. It was comfortable, though sterile, compared to my wood-mud-water house. The rain silently drained from the cold, dark clouds outside—clouds I could watch from this secure environment like a movie, rather than working under them in the carpentry project. Finally Mr. Brown arrived and asked my business.

"Mr. Clarneau?" he said, half smiling, half solemn, warm in a noncommittal way, like the architecture surrounding us. I stood up and we introduced ourselves.

"Pleased to meet you." he said. He was far warmer than his secretary, treating me as a client, with respect and without the cold, critical air.

"Well, Bill, what can I do for you?" he asked. He conveyed so much warmth it sounded as if he were saying: "okay, what

42

problem can we solve for you today?" He was confident, poised to excise any trouble, large or small, with ease.

"I need a construction loan," I said. "I'm building a house. It's sixty percent complete and I'm the sole owner. But I need money to finish it." As I understood loans, as long as a person has something of value equal to the loan amount, he can get a loan by giving the bank ownership of the property until the loan is paid off; if he doesn't pay off the loan, the bank can keep the property, sell it, and receive the loan amount plus any profit. I thought the half-completed house could act in this regard; it was worth far more in its present state than the amount I wanted to borrow. I planned to mortgage the house to get the loan. On my last bank visit, I'd had trouble because I wasn't a contractor, but I thought I had solved that difficulty.

"Are you a contractor?" he asked. I'd discovered that no tests, school, or experience of any kind was necessary to become a contractor—only a cash bond.

"No, but I could become one by posting a bond," I said. He nodded, satisfied. He asked several things about the house—costs, methods, materials—and sounded increasingly positive; I thought he was going to grant the loan. But then the roof fell in.

"Bill, what sort of work do you do?" he asked. To me the answer was apparent, the house was my present occupation. I had experience in several other fields, but none was more solidly respected by society, in my thinking, than the present occupation. I was by now a competent contractor.

"I'm a builder," I said. I had even been thinking seriously about building another house after this one, provided it didn't take too long to finish my present work. I said it with confidence. His smile retreated.

"Do you have some other property for your next house?"

"No." I shook my head. His features changed. New lines appeared on his face. The room teetered slightly.

"Well, frankly, Bill, I don't see how you're planning to make the payments on this loan. How are you going to build another house? If you're a builder, how do you plan to launch yourself?" How could he be so insensitive?

"When the house is done, we'll borrow against it for the whole amount it's worth," I told him. A larger mortgage would then give us the cash to pay it off—provided the house was sold within the next few years. It all seemed simple to me. He shook his head again.

"No, we can't do business like that." This flat response infuriated me. Their tradition was going to stand in the way of my house being built!

"Why not?" I asked.

"We just can't. It's poor business. We can't lend money to pay off other loans."

"I don't see why," I said. "Where do other builders get money, if not through construction loans from you? If the loan doesn't work out you can repossess the house." I was sounding desperate. I was desperate.

"Most builders have their own source of money. We make construction loans to contractors only through the home owner, never directly. Some builders have a good credit standing with us and we make personal loans to them. We have to know them very well first. We can't do that for just anyone." His position was firm. I still had another approach.

"In that case," I said, "we can certainly meet any payments for a loan this small with my wife's salary—even if I never work." This logic was, to me, compelling. But he recoiled on hearing the last three words.

"We can't consider your wife's income in evaluating your ability to pay." I was dumfounded! "We consider it as additional income along with the husband's. But it would never pass the loan committee this way."

I had worked so hard for the previous six months, not simply 40 hours a week, as a normal job would have meant, but 70 or more. I'd done everything very well by professional standards, but this man was telling me now that my labor was worthless. And worse, that Diane's was too—and she had an established job! A call to her school would prove to him that she was highly respected by her associates, a stable person and an excellent teacher. But no, the bank was deaf and dumb.

"The only thing I can recommend," he said, "is for you to get a job and work on the house in your spare time. After a while you can come back in and we'll be able to look at the situation again."

I could feel the room beginning to spin in the silence that followed. If I got a job the house would never get done in my "spare time." I was about to go crazy anyway. It had already taken so much longer to finish the job than I had expected. What sort of an institution was this that would overlook my proven ability so completely? How could this company be so insensitive to reason? Even if I got a job and remained there for a while, didn't he see how poor a guarantee it would be for the bank—that I could quit the day after the loan was made?

"Have you tried other banks?" he asked.

No, I hadn't. I didn't think going elsewhere would help, either; I could hear the next official saying, "Well, if your own bank won't lend you anything, don't expect us to." My head reeled, the walls were closing in. I had to get more money to finish the house. The house was worth a lot, but no one cared. I couldn't leave until I'd exhausted every last appeal.

"No," I said, "I was hoping that the bank I deal with—the one where I keep my account—would be of help." He wasn't moved.

"I just don't see any way I can help you," he said. "Even if I did, I know the loan committee wouldn't pass it. I *know* that."

"Would it make any difference if I were simply on a leave of absence from a job?"

"It would be better, I think, even if you were collecting unemployment," he said.

I could hardly bear the irony.

"Can you think of any way I might get a loan? Any way at all?" I asked. He thought for a moment.

"Well, if you could somehow finance the house to completion—maybe get the lumberyard to give you credit—it would be easier for us to look at it after it was done. But then you might as well sell it yourself." We were adversaries now; he didn't trust me, I was too unorthodox. He reached into his bag of cliches. "I don't think, now that I know your situation, that we would touch it with a 10-foot pole," he said. "The way we work here is strictly on residential real estate. We don't make any business loans. You want to start a business. We can't help you with that unless you have an insured way to pay us back, independent of your business. I think you'll just have to get a job. I don't see any other way."

The conversation was ended, but I couldn't face the fact. I pleaded.

"Look," I said, "I'll do anything you want. I'll try getting the lumberyard to extend some credit. My wife has a very stable job. We're both hardworking people. Isn't it clear that after doing all that work I wouldn't ever let a payment go by? Isn't there any way?" My speech moved him slightly. He thought for a moment.

"I just don't see any way," he said finally. "I can't help you." I thanked him for his time. He'd given me the rundown of how banks work, but had his hands tied by "his superiors."

I arrived home in a nervous fit. I had no faith at all left in banks. I was dizzy from this untimely setback. There was nothing I could do. I didn't think anything would work. The house would stay an empty shell, unlivable and useless.

My feeling for the house was changing. I began building the house with the thought that I could complete it myself—no loan, no mortgage, no monthly rent to pay. Everything up to this time was flavored with thoughts of owning the whole thing when it was finished, of making my sacrifices pay off by having free rent afterwards.

BACK TO THE BANK

Early the next morning, I went to another bank with a revised story. I told them I was building my own house, needed some more money, and that I would return to my work in advertising when I finished. I would competently finish the job; they could view my do-it-yourself, amateur expertise by visiting the house. I'd be happy to get a contractor's bond if necessary. I admitted I didn't have any line on returning to my old job, but I exuded confidence that somehow I would have no trouble finding another position.

They took a copy of the house plans. I filled out several forms and spoke to an officer, repeating my story. I mentioned my wife's income and it carried more weight. Things seemed positive. Three weeks later we got the loan.

We would have rejoiced, but by then the house was simply a war against time and the depression of my damaged hopes. I thankfully accepted the loan. Diane and I signed the mortgage.

"This mortgage," read the official stationary, "matures June 7, 1998."

PERMITS AND LICENSES

When the new plans were finished, I could visit city hall to get the *building permit*—or so I thought. I was getting weary of the preliminary paperwork, anxious to start the house. Even though I had worked in government offices before, I was frightened on my first visit to an institution that held the power of life and death over my fledgling house. I walked into the *buildings* room and, after standing around for several minutes, discovered a sign that read: PLEASE SEE THE RECEPTIONIST. I saw the receptionist and stammered that I was going to build a house and wanted a permit. She waved me to a waiting area, and said someone would be out as soon as he was free.

Afternoons are the worst time to conduct business at the city hall. After an hour, the delay helped turn my timidity to angered frustration. Finally, a square-featured man in his forties, with a face marked from many years of haggling over bureaucratic nonsense, appeared behind the counter.

"Okay, who's next here!" he barked.

I jumped up and put the plans on the counter. He spread them out and quickly flipped through the pages.

"I'll need three copies of each page for the permit," he said. "Fill out this form." I did so and, after a half-hour wait, returned to the counter. "Okay, let's see if the lot is legal." He opened the zoning map book, found the lot number, and checked the zoning rules from another text. "Square feet...yep...frontage...yep. Looks okay." Noticing the engineer's name on the plans, he said, "Oh, he's a good man. You won't have any trouble with these. Now all you need are two more copies and $56.21. These are taking about three weeks to process, so you can pay then if you'd rather."

His last statement came as a shock. I wanted to begin work that day. This three-week delay was something any contractor would have planned for, and was something I had no choice but to accept. I thanked him, said I'd be back shortly with the other copies, and left for the engineer's office to have the copies made. As I turned to go, he said, "If you want to check on your sewer and get a house number, you can do that now."

I'll never get out of here, I thought. He pointed to the elevator, and handed me another form.

"Fourth floor," he said.

After another hour of waiting, filling in forms, and watching a clerk verify that my lot had a sewer available as required by law, I returned with all the vital details completed. I handed the papers to the receptionist and departed for home, plans in one hand and a manila envelope containing a card with my standard city house number in the other. By now it was late afternoon. I clutched the numbers—the first tangible evidence that the project was under way.

BACK TO CITY HALL

Several weeks after applying for the building permit, I called city hall to check its progress, hoping it would be ready early. I asked to speak with the man I had dealt with earlier. The receptionist said he was on vacation, that another person had made a preliminary check (indeed, at last I was rid of that drill sergeant), and that he had found a list of deficiencies I would have to correct. How could this happen if my engineer was such a good man? I wondered. I retrieved my plans and spent the whole weekend, day and night, correcting them according to the instructions left by the second man.

Returning first thing Monday morning, I asked to see the man who made the list, so I could show him how the problems were corrected. The receptionist said that the original man was back and, since the second man was busy, I could speak with him.

It was early morning this time, which evidently helped, since he came right out and examined my plan. He surveyed the checklist, looked back at the plan, and spoke under his breath.

"Oh, Mr. Jackson's a *commercial* building-plan checker. You didn't have to go to all that trouble. Here, I'll just give you the permit. Have you got fifty-six-twenty-one?" I did, and eagerly wrote out the check; he gave me the permit.

As I was leaving, I noticed a bulletin board where applications for building permits were posted, according to the date they were submitted; my number had somehow fallen behind its proper sequence. I concluded that regardless of how long they said it would take, to make things go faster I should have called more often to check on it. I also realized that the amount of trouble the bureaucracy can cause depends greatly on who you see. One anonymous checker had threatened to scuttle my plans; another had, in spite of his tough facade, cut redtape to help it.

On the permit was a note directing me to the plumbing bureau. By this time the bureaucracy of city hall lost all of its threatening aura. I walked into the plumbing office, expecting to fill out a routine form. On this particular day I had been clearing the lot, and was a dirty, sweaty mess, a striking contrast to the neatly dressed contractors and architects city hall employees are used to seeing.

The man behind the counter asked my business. I told him that the buildings bureau had sent me, regarding the storm drain, as indicated on the note. He replied that I wouldn't have to worry about that because it was the plumber's business—who was the plumber? I was, I said. He looked me up and down, silent for a moment.

"You gonna cooperate?" he asked. "Cause if you're not gonna cooperate, you won't get the permit."

"I'll cooperate. I'll cooperate," I said, irritated at another governmental stuffed-shirt.

"We've had too many guys like you lately that didn't cooperate," he said, with an energy fueled by something I couldn't see.

"Look, I'm building a house—my own house—I planned to do the plumbing myself—"

"*You* can't do it yourself," he said emphatically.

"How do you know that?" I said as forcefully as I could, thinking maybe he was right, as so many of my relatives had thought. He backed off slightly.

"Well, you'll have to show us you can do it before we'll give you the permit." That sounded a little fairer.

His behavior was totally unexpected. Although I knew plenty of people who had done their own plumbing, I thought maybe he could pull out some obscure rule to keep me from doing mine. I didn't realize that the law was on my side, that anyone can install plumbing in his or her own house, and that he had to give me a permit. His face glowed with scorn, though.

"Okay, what do you want me to do," I asked. With this statement, the bureaucrat could see that he had won; he softened further.

"Well, just dig that *french drain*." He pointed to the paper I'd brought in. He explained that the drain had to be 40 feet long, 3 feet wide, and 3 feet deep. Not knowing how much work it would take, I agreed, happy to hurdle this obstacle.

"Can I have the permit when that's done?" I asked.

"When it's done, we'll have our man come look at it, and see if you did things right. Then we'll see." I decided in my best schoolboy manner, then, to do a perfect job on the drain, so they couldn't refuse the permit.

He introduced me to the inspector who would come out, who in turn gave me his card, saying to call him when I was ready. I thanked them and left.

With the permit finally issued, we celebrated. It was my diploma from the buildings bureau; and, instead of the bland statement about having the right to build a house, it should have read: "We hereby certify that William Clarneau has successfully completed the required course of rigmarole this day, August 18, 1974."

Buying Materials

Even though I couldn't legally begin the house until the permit was granted, I could clear the ground that the house would sit on. For the next three weeks, I hacked at the underbrush with a machete, sawed the few maple saplings in the way (I positioned the house in a place where no large trees had to be felled), and hired a surveyor to mark the corners of the house in the correct places on the lot. (This is requirement for houses located within 10 feet of a property line.) Then I made the marks where the foundation holes were to be.

COMPARING BIDS

In addition to the ground clearing, I started looking for materials. As with buying the land, I was at first ignorant of how prices for building materials worked. I didn't think a person bargained for them as he did for land, but I was sure that someone building something as costly as a house couldn't pay retail prices. From a job I once had in the furniture business, I knew some goods could be bought for 60% less than retail. Hopefully, lumber prices would work similarly, and I could find a wholesaler that would sell what I needed.

I paid another visit to the old family friend who was a contractor.

"Well, *Pack Lumber Company* is just about as good as anywhere," he said. "I've been dealing with them for over twenty years, and they're about as cheap as anybody." I recognized this company as a retail store. It puzzled me that a contractor would buy there.

"Do they give you a contractor's discount or something?" I asked.

"Oh, sure. Just tell 'em I sent you. They'll fix you up. Just tell 'em I sent you."

I wanted to see what a person "off the street" would encounter at the lumberyard, however; I didn't want to use his name right off. First I wanted to see if any other dealers might have better prices, in spite of the contractor's long-term experience.

I didn't know the first thing about lumber at that point; I couldn't tell a two-by-four from a two-by-six without measuring. I has visions of walking into a store and displaying my ignorance so well that the clerks would feel safe in charging me double for everything, knowing I would never find out. I had nightmares about wasting so much money buying lumber that the house couldn't be finished. Clearly, I had to protect myself; I decided to visit every lumberyard in the area, asking each one to give me a price for my lumber needs. I took my plans to 12 stores altogether.

I had been in lumber stores no more than three times in my life prior to that time, always finding a gruff weatherbeaten clerk behind the counter, and several outdoorsy, flannel-shirted, do-it-yourself types milling about. I never felt at ease there, as I do in supermarkets, for instance.

I threaded my way through the domesticated lumberjacks and up to the counter. Painfully self-conscious, aware that my ignorance was glowing like a neon sign without my even saying anything, I knew everyone was staring at me. I fumbled out some squeaky words to the clerk behind the counter.

"I'm building my own house," I said weakly. Surely I heard snickers. "Can you take these plans," I said, putting the plans on the counter, "and figure out the lumber and the price?"

"You want me to do a takeoff on those and give you a price?" was the reply. I nodded. "You'll have to leave them for a few days. I can't get to it until then. That all right?" I nodded again. "Okay, come on back to the office and I'll get your name and address."

I followed him back through the yard to his office, where he wrote down the vital facts. I had several more stops to make, but was so relieved to find the clerk talkative—a friend in this hostile area—that I wound up listening to a complete history of his recent surgery. I relaxed; anyone who would collar a stranger to listen to his troubles was less than a case-hardened mean old lumberjack; lumberyard clerks are human too, I decided. But the next time I came I promised myself, I'd grab the estimate and run.

The next place I visited had fewer threatening figures standing around. I submitted the plans, received a similar response, and agreed to return in a few days. I drove along, with my fear subsided; I was in business now. I exulted in this new role, complimenting myself on my shrewd business sense in getting these bids.

At the following stop, I presented the plans with the same explanation. The surrounding people again seemed to be looking at me peculiarly, wondering, I thought, what this stranger was doing in their store. When I mentioned the address where the lumber would be delivered, the clerk hesitated.

"That's inside the city, isn't it?" This lumberyard was some distance away from downtown, in a suburban area. I replied that it was indeed in the city, but they delivered there didn't they? "What are you doing?" he asked, "trying to get competitive bids?" His tone was accusing, but I had done nothing wrong that I could see, unless my ignorance had somehow foiled me after all. Of course I wanted competitive bids; wasn't that what any good businessman wanted?

"Yes, it is," I answered.

"Well, I can't make one from these plans," he said.

"Why not?" I was getting irritated. The other dealers had agreed to accept my plans and quote a price. He would certainly accept plans from one of the "regulars" standing around the counter, wouldn't he? What was he doing? Just giving me trouble because I looked like a kid? He wouldn't treat me that way! I'd go where I was treated like a regular customer and a human being.

"You can't get a competitive bid from someone else's takeoff sheet," he said. "A lot of guys'll leave something out. If you want me to bid against another yard, I'll have to have a materials list. Then, with all the items the same, you can compare prices. Otherwise, a guy can leave something out to make the total price lower, even though all the items cost more." My anger melted into a hot pool of embarrassment. I was a genuine ignorant kid after all. "Do you have a materials list?" he asked.

"No...Well, I *have* one," I said, meaning the one I bought from the plan service, which didn't apply to the modified house, "but it's not complete."

"Well, you bring in a materials list, and I'll be happy to bid it for you," he said, turning to the next customer.

"Okay, I'll bring one in." I left, wondering how I could make a materials list. I couldn't do a takeoff, as they called it—a list of the lumber needed in a house plan. I had only the vaguest idea of what comprised a house; I couldn't figure out the lumber needed for the skeleton, much less the whole house. I was headed toward home now, wondering what to do. I decided to visit a lumberyard recommended by my builder friend, where they supposedly didn't mind helping an inexperienced person.

Arriving there, I was ushered to a back room, where the "plan expert" worked. He was the most cordial of anyone I had met thus far, and he said he'd gladly do a takeoff on my plan if I'd give him a few days. Although that left me back where I had been at the other stores, I realized that if I waited until he finished with the plan, I could compare the takeoff with the plan and hopefully decipher how it was done.

On my way back to my building lot, I passed a construction project—a warehouse. A carpenter was building wooden forms for concrete. Although, by this time, I had read some books on how to frame a house, it seemed like a good idea to see the real thing in progress. I parked the car and watched him working on the form.

A young man not much my senior, he wore a green T-shirt, blue jeans. The thing that distinguished him most of all as a professional—a carpenter's leather tool and nail pouch.

Casually, he hammered here and there. Then I watched as he played with what looked like a string and a level. Soon I walked over, announcing myself.

"I'm building my own house, all by myself, and I don't have any experience. Are there any special tricks to what you're doing? I've never worked with concrete before." He was very nice, sharing my excitement about my project. (Since then I've concluded that many carpenters dream of building their own house.) He showed me things called *form ties*, which I didn't know about, since I'd been reading outdated building books. He showed me how the form ends are secured, pointing to the form.

"Right here," he said, "at eighteen inches from the bottom is where the pressure gets concentrated, so you have to make sure the forms are good and solid here especially." Another man appeared—evidently his superior. The carpenter looked up at him. "Hey, Jerry," he said, "can you think of any special hints or tricks to give this guy for making forms? He's building his own house." The man walked to the form and pointed to the 18-inch level.

"Gotta make sure you get it real strong," he said.

"I told him that," said the carpenter. "Can you think of anything else?"

"No. Not much to it, really. Just make sure the form is strong so it won't break when the concrete goes in."

I pointed to the footing section, to a part where it looked as though the wet concrete would run out through the bottom and thus keep the form from being filled.

'Won't it run out the bottom, there," I asked.

"Oh, no," they both answered, "the stuff's too thick to run up over there."

"Jerry's building his own house, too," the carpenter said. "Maybe he can help you with something else." The boss thought for a moment.

"Yeh—ah, hey! If you need any heavy timbers—do you need any big beams or anything?" "Maybe for the roof," I said.

"Well, there's this guy down at the old Timber Maker's building—they're wrecking it now—who's selling used stuff, good stuff, for almost nothing." He gave me the address. "Just ask for Mike. He's got other stuff, too: plumbing, doors, windows, and some smaller lumber. But the big beams are especially good. I was there last weekend, though, and most of it may be sold by now. You can try it and see what happens."

I hadn't, until now, given any thought to used materials. I didn't know what to do with new material, and had little use for further complications. But cheap materials sounded attractive, especially after seeing the first sample estimate on new lumber.

Remembering the forms, I asked when they planned to pour the concrete and if I could watch. I could tomorrow, they said. I thanked them and left, elated at my information-filled stopover.

Since the day of bid-collecting was upset anyway, I decided to visit this building wrecker, to see what he had left. The address was in an unfamiliar area of the city, so I called the wrecking company to verify the location; after an hour or so of being lost, I found the half-demolished building.

The area looked as if a bomb had been dropped on it. Debris and a couple of rusted and wrecked old pickup trucks littered the area. The place looked deserted. But when I got out of the car, I heard voices. A fence surrounded the area, with signs at regular intervals reading POSITIVELY NO ADMITTANCE. I hesitated, thinking maybe I needed a guide through the debris to avoid having anything fall on me. Seeing no one, though, I timidly proceeded toward the voices, and peeked inside what remained of the building.

Several men, chainsaws in hand, were perched here and there like giant vultures feeding on bits of the roof and wreckage. A dumpy, stubble-bearded man wearing a hardhat was stomping about, giving orders to the others as if this were the Queen Mary and he was her captain. Just as I approached him, the chainsaws chimed in full blast; I yelled that I was looking for Mike. He motioned toward the entrance where we might hear each other better. We reached the outside area, where the noise of the saws was muffled.

"I'm looking for someone named Mike," I said.

"I'm Mike," he said with friendly tone, "What are you looking for?" I still didn't know, since I hadn't a materials list, but I had a couple of things in mind. "Need some shiplap?" He pointed to an enormous pile of what I thought was just trash, boards piled every which way, full of nails, broken in many places. "Beautiful stuff," he said, nodding his head with certainty. "You can have that for nothing—just pull a few nails—we have to get it out of here."

"No, I don't think so," I said. I had no desire to acquire garbage—it looked terrible. "Have you got any six-by-six?" I asked. I knew I needed some for the foundation stilts. He slapped his hand on the post I was leaning on.

"There's your six-by-six," he said proudly. "Be a while 'fore we get it out, though. You need it right away?" After my land bargaining success, I thought I would resurrect my tough businessman/bargainer facade.

"I don't know," I said. "How much do you want for it?"

"Seventy dollars a thousand," he said. (By then I knew that lumber was priced so much per thousand board feet—a *board foot* being equal to a piece of wood a foot wide, a foot long, and an inch thick. The price sounded very low, but I wasn't sure.) "Beautiful stuff here," he said, pointing to the post I was leaning on. "Can't buy it new like this now. Look how bright it is." He pointed to some golden pieces in the ceiling. I wondered if he was trying to take advantage of me with his sales pitch. I didn't answer.

"What else have you got," I asked. "Any four-by-six?" I was thinking of the roof beams that I would need.

"No. We had some, but we moved it out already." He made it sound like a giant corporation was "moving the goods." I looked around this gargantuan rat's nest, amused at his seriousness. He showed me several other trash piles, then one of his chainsaw operators called, and he excused himself. "Well if you need some shiplap, we got plenty of good stuff left here," he said, and returned to the building.

It appeared to me that he was out of anything I needed. But I was awakened to the possibility of saving money with used materials. As soon as I got home, I went through the telephone directory, and called all the building wreckers in the city, asking each if they knew of any salvage places. I learned about only one—a used window, door, and plumbing supply store. I called and spoke to a woman who said they were open in the afternoons, but if I came out tomorrow morning she would be there. It was a long distance to•her store, but I was enjoying the fantasy of getting house materials for free or nearly free, so I decided to drive out there to investigate.

Diane and I went together, our heads filled with visions of a storehouse packed with stained-glass windows, hand-carved doors, antique plumbing fixtures, and the like. But all we found was a run-down building with peeling paint. Just as the woman had directed, it was one block beyond the "No Return" tavern, closed in on three sides by a sprawling auto wrecking yard.

Toilets, bathtubs, pipes, windows, doors, and the proverbial kitchen sink lay in clumps around the building; with seeds growing out here and there as if arranged by some avant-garde designer. Perhaps the inside would be more refined, we thought, and marched up to the entrance. Several CLOSED signs were pasted on the window; we thought they might be antiques themselves and peered inside, knocking and waiting for the voice from the telephone to materialize. No one came to the door, however, so we window-shopped.

We were disappointed to find that no great treasures were evident; the room was jam-packed with old doors, windows, and more plumbing fixtures, reminiscent of Mike's trash piles. Had something from our hopeful visions been there, we might have waited a while longer; but knowing nothing about how to tell a good old door or window from a useless one, we left.

I let the idea of buying used materials drop for a while, at least until I knew more definitely what I needed. After several days of digging and clearing the lot, I returned to the lumberyards where I had left copies of my plans.

First, I visited the man who had been so nice. I asked him how he'd figured out the lumber, and he said something like, "Oh, you just start at the bottom and work up." Seeing I didn't understand, he added, "It's easy. Just follow the list and you'll see what I mean." I took it home and did so; he was right, but I had to call him a few times to see how he had done this or that. I took his list, with a few additions, and made copies of it. At last I could distribute it to the other lumberyards.

With all the estimates collected, it turned out that the dealer who gave me the first bid had the lowest prices after all. But, surprisingly, the estimates varied only slightly from each other, in relation to the retail price. Some stores charge as much as 10% less than retail—but none was anything like the 60% I hoped for. I wasted a lot of time running to all those places for such a lack of satisfying results, although I knew now for certain how lumber prices worked, and had also learned how to do a *takeoff* in the process.

Despite my diligence in seeking competitive prices, the estimates were more than double what I had expected—quite a shock, considering my limited funds. Again my housebuilder friend had led me astray; his comment that a house could be built for only 25% of the finished value wasn't worth any more than his saggy-walled house. I panicked at first; the price was so high that I shouldn't go ahead—yet the land had been purchased and partially cleared, with many hours of my own toil.

Maybe I *could* find most of my material needs in used lumber. If I could but manage to distinguish a usable piece of material from a useless one, maybe things would work out. I regained my composure, thinking maybe a combination of used and new materials would make up the difference between the present problem and my original plans. If it meant the difference between completing this project and not, I wouldn't mind pulling a few nails. I remembered Mike's trash pile; it was free. If I could get all the wall sheathing out of that junk pile, that would save $300 alone.

I didn't have the building permit yet, and the foundation still had to be finished. Perhaps by the time I was ready for it, the lumber would have dropped in price from the present all-time high. Perhaps my friend hadn't led me astray after all and would, when he returned to town, point out a few more places where I could save money. He was, after all, the only one I knew who had ever built a house single-handedly. Maybe that fact alone resulted in money savings that normal contractors never obtain. Meanwhile, I would look for used materials and do the work at hand.

BUYING LUMBER

You can save money by taking advantage of the fact that wood is an amazingly versatile and reusable material. A board could be first used as part of the foundation form. When the forms are dismantled, the same piece of wood could be part of a scaffold. When the scaffold is no longer needed, the same piece could find its final resting place as part of the house itself. One precaution for reusing lumber: A bit of cement or a nail will dull a sawblade instantly. Unless you're careful not to hit such things, the money saved reusing lumber will go to buy new saw blades.

New Lumber

Before I started working on the house, I thought of wood as a single substance. But wood varies from one tree species to another; each has its own odor, weight, and hardness. Cedar and redwood are quite fragrant, and they are light and soft; they are good for insulation and room paneling. Douglas fir is soft, medium weight, and durable; it's used for house framing. Oak or other hardwood is almost impossible to drive nails into without splitting, and is mostly used for making furniture.

Lumber comes in certain basic thicknesses and widths, with various standard lengths available in each. A 2- by 8-inch board is one of these. Simple enough? No. Just to confuse things, 2- by 8-inch board (called a *two-by-eight*) isn't really 2 inches thick and 8 inches wide; it's slightly less: $1\,^9/_{16}$ by $7\,^{13}/_{16}$ inches. Two by eight inches is the *nominal* (or named) measure, the measure of the board when it's first cut from the log and comes out with ragged, splintery edges. Its *actual* size is called the *dressed* or finished dimension, the size it is when you get it at the lumberyard—smooth and ready to use. The floor will have edgeboards $1\,^9/_{16}$ by $7\,^{13}/_{16}$ inches, even though you call them two-by-eights.

Prices for lumber are quoted in *board feet*. As I said, a board foot is an inch-thick piece of wood a foot wide and a foot long. A 10-foot-long four-by-twelve, for instance, is nominally 4 inches thick, 1 foot wide, and 10 feet long: $10 \times 4 \times 12 = 40$ board feet. In board measure, then, a four-by-eight equals two two-by-eights. Since you pay according to board feet, a four-by-eight should be exactly twice as expensive as a two-by-eight for equal lengths; thicker lumber, though, costs more for each board foot.

Lumber also comes in different *grades*, according to the section of the tree the piece comes from, which determines the board's strength. Most house parts must be made of *standard* and *better* grade, also called "number two." You can simply order by name from the store, trusting the employees to pick the right kind; or you can visit a store and ask to see some samples of the different grades.

Used Lumber

It's a fact that our society is incredibly wasteful; you may be able to profit from this fact by using the waste. The building industry, for several reasons—tax advantages, building codes, graft—likes to build new buildings rather than repair old ones. Thus a lot of leftovers are available from demolished buildings just for the cost of removal, or for a small salvage contract. When freeways are built, the displaced houses are sometimes given away (or sold for no more than several hundred dollars) to anyone who will remove them. For someone with little money, scrounging for used parts can bring wonderful results. The parts were already paid for when they were new; their only cost is the labor it takes to remove them.

Used lumber can be had for as little as $70 per thousand board feet. Prehung doors and windows for just a few dollars are usual. Used toilets and sinks for $2 each (versus $35 each new) are also possible.

Interior decorators sometimes change people's carpet and discard the still-good carpet. Building-materials stores are going out of business all the time and sell things at a loss, or at least *at cost*. Newspaper advertisements are a good source of information on used materials.

Lumber stores sometimes have display doors and windows with nicks or other minor damage that makes them hard to sell, and will let you have them below cost. Window or door factories sometimes have warehouse sales with discontinued or hard-to-sell items they're glad to be rid of. Occasionally, they have products sitting around that were ordered but never picked up, items they will want to almost give away. Almost anything but nails is available used; plywood, roof tile, furnaces, plumbing supplies, kitchen appliances, bricks, lumber, etc.

NEW OR USED?

Used lumber is especially applicable during the framing phase of construction, since everything will be covered by the finished work. It makes no difference if paint and old nail holes remain. Used lumber, however, has drawbacks you should be aware of. It's very dry and brittle, and splits easily unless the lumber is three or more inches thick. Dryness has another drawback, regardless of size: it's harder to drive nails into old lumber than into new lumber.

New lumber is still full of sap and moisture; it's much softer, and nails usually go right in with a few hammer blows. Kiln-dried (dried in ovens at the lumber mill) lumber has this problem also, although not as pronounced. Driving nails in dry lumber is murder for anyone but a practiced carpenter. The roof of my house is all kiln-dried lumber, and I averaged four or five nails bent to every one driven in. This job took three weeks of utter frustration, and could have been easily completed in half that time with green lumber.

One advantage of dry lumber is that it doesn't shrink. Green lumber shrinks like old-fashioned jeans, a little at a time as it dries. The lumber industry says a 2-by-8, for instance, will shrink to 1½ by 7¾ inches. (Douglas fir doesn't shrink lengthwise.)

Since my roof is exposed to the inside, I didn't want to have little cracks appearing between boards as the lumber dried. I used dry lumber for the roof. A house built with new lumber will settle a small amount as the lumber shrinks. Factory-dried lumber is very expensive and hard to use, however, so no one used anything but green lumber for framing. Incidentally, new lumber arrives with the ends painted, sometimes green, to protect the wood from splitting as it dries—not, as I once thought, to indicate whether the wood was dry or green.

All lumber will crack and split as it dries, however. Dry wood (new or used) is good for exposed parts, especially beams, since the cracks are already present when you obtain the lumber; they can be filled in with putty and hidden. With new green lumber, cracks will appear later, after the house is complete—when it will be messy or more difficult to apply putty or repair the exposed surface.

I acquired some used 1- by 8-inch boards to use, instead of plywood, for the floorboarding; and, although the price was one-fourth that of plywood, I had to remove thousands of nails before it could be used, pick up and deliver it myself, and apply it piece by piece, which took far longer than plywood. Many pieces split and had to be discarded. Plywood would have been much faster to install, and more economical as well. I purchased enough to sheath the walls, too; but after seeing how slowly it went on the floor, five times longer than plywood, I decided to use plywood on the walls.

Another problem with used materials is that even with a reliable source, it's seldom possible to obtain what you need. You can't walk in and buy whatever you want; you have to take whatever is there. Such a process can waste a lot of time, running back and forth to the supplier. With new lumber you can call the store and have whatever you need delivered. With long lengths of lumber, too, it's hard to transport unless you own a flatbed truck. Only if you can get everything at once can you justify renting a large truck to transport materials cheaply.

DIGGING AND MORE MATERIALS

I still had much to do before the house could really begin. In addition to the plumbing assignment, the hillside site had a tiny spring running through it, necessitating a mammoth drainage system encircling the house. To make the drainage system, I had to dig a ditch, fill it halfway with gravel, lay some pipe, and fill the rest of the cavity with more gravel. Including the french drain, then, I had to dig 240 feet of ditch and put in the pipe and gravel.

What can I say to describe how hard that was? It took three weeks of 12-hour days, including weekends with friends helping, to complete the wretched system. I came home each night with muscles twitching uncontrollably from the strenuous overwork. The work was utterly boring, hard, and unrewarding; and I hated it. After it was done and covered up, it was as though the three weeks had been spent in some far-off prison.

The french drain alone was an experience in itself. Diane and I found ourselves, one blistering afternoon, standing in the half-dug excavation in the concrete-hard sun-baked clay, sweating and almost delerious. All of a sudden we ran to opposite ends of this grave-like opening, turned as if to fight a duel, and broke into a slap-happy rendition of "If I Were a Rich Man" from *Fiddler on the Roof*. Dancing, capering back and forth, we imagined ourselves slogging through our earthly toil, happy peasants, forever doomed to dig this awful, pointless hole, waiting for the film crew to catch this act of spontaneity. Stumbling up and down the ditch, we strummed our shovels for guitars and until we fell, roaring and insensible; our task was hopeless, the situation absurd.

That was one of the few "high" points, though. Bringing gravel down the 150 feet of narrow, twisting, and steep path in a wheelbarrow was downright hair-raising. Each time I made the trip, the load, weighing more than myself, would strain to get away under gravity's coaxing, while I dragged my feet, skidding along to hold it back. Diane, who couldn't even push the empty wheelbarrow back up to the road, watched and oversaw the gallery of onlookers who were convinced of my insanity as I skidded off down the path for another run.

It seemed like there must be an easier way to move materials than by wheelbarrow. I tried to make a pulley system between several trees. After spending $50 and a week's time perfecting it, the rope stretched so badly that the load acted like a slug, hugging the ground as it went, as if on a leash instead of the pulley. I was done with mechanical devices after that fiasco—back to the wheelbarrow and brute strength.

Finishing the drainage system, however, revealed yet more preliminary work that I managed to overlook, in hopes that the building permit would magically sweep it away: The 300-foot sewer line had yet to be laid; the foundation holes had to be dug and forms had to be made before the visible house could even begin; and I still had to find some cheap materials.

I had the idea that I could complete the house in two months; but with six weeks already gone, I had to revise my thinking and change my plans for the coming year.

I didn't want to spend a lot of time building this house, yet I didn't see how I could avoid it. I was terribly depressed. Digging and other hard labor was making me sick, and I had only more of it to look forward to in the coming weeks. If only I could quit. If only I could find someone who could get me out of this mess. What could I do? I was surely going to lose my mind at this rate—and I hadn't even *started* the house. I despaired. I was lost in a nausea of self-pity for getting myself into this worst of all commitments. I didn't know what to do. I had invested so much money and hard work already that I had to continue; but it made me sick to think of how much yet remained to be done. Surely something awful would happen before this was over.

I decided to have someone else build the concrete forms and pour the cement. I contacted several places about the concrete, and planned to have one do the work in three weeks, the earliest anyone could get to it in the busy summer season. I figured I could excavate the day before with power tools, and went to work finding cheap materials in the meantime.

The thought of Wrecker Mike's trash pile had improved with time and I returned to take advantage of his bargains. He was still there, king of the salvagers, but the pile I wanted was gone.

"Do you have any more shiplap?" I asked.

"How much do you want?" he said. I told him the amount indicated on my estimate sheets. "I've got just the stuff for ya," he said, beckoning me to trail behind him, into the remaining portion of the wrecked building. "Seventy dollars a thousand for this. Beautiful stuff," he said, as we reached some men pulling up flooring. He picked up a piece. "Beautiful stuff." He sounded like a recording from the last sales pitch he'd given me. "Can't buy this any more. Look how bright it is. Beautiful stuff!"

I still couldn't tell whether or not he was selling me trash. But I couldn't see anything wrong with these boards, either.

"What happened to the free stuff?" I asked. "I thought I could get this free."

"Oh, not this," he said. "That was junk—got taken over the weekend." He waited for my reply. His enormous potbelly hung there waiting, too. His hardhat looked like a toy on his large, bald, bowling-ball head. I pointed to the boards below.

"This is one-by-eight, right?" I asked.

"Yeh," he said, motioning to another pile of boards that looked identical to me. "Here's some one-by-six. Can you use that for anything? I'll give you a good price on that—you bein' a businessman and all—fifty dollars a thousand for that." It looked okay to me, so I agreed to buy it. "Okay, we'll load it up." He yelled to one of the men prying up boards.

"Have you got any two-by-fours?" I asked. He said he'd look, he had a few downstairs. We went down to look while they loaded the long, thin, nail-filled boards onto my rickety pickup truck. Mike stood there in the heat, watching his men sweat, picking out broken boards to toss them aside, and okaying the several two-by-fours that got thrown on. When the pile of boards was transferred to the truck, he said he'd have the rest ready tomorrow. I was disappointed, since I'd thought I could get it all at once, but agreed to return.

It was a hot summer day. I smelled the asphalt "cooking" in the sun. I felt sluggish. My truck was sluggish too, I guess. It refused to exceed 10 mph, no matter what I did to coax it faster. It took over an hour to return to the lot, where a friend and I haphazardly stacked the lumber on the shoulder of the road, throwing it from the truck to the ground.

The following day, though, Mike didn't have the next batch ready for us.

"Ah, these guys you get workin' for ya," he said, "don't show up for work. It's gettin' so I can't get help. A man might as well give up and retire. I'm sorry, but I don't have your order ready. Try again tomorrow. If they're not here then, I won't hire 'em again." He was fuming. We gave him our condolences, said we'd be back, and left feeling as though our project was slowly grinding to a halt.

Back at the lot, we started pulling nails from the boards—a task not as hard as digging, but incredibly boring, and hard on the back from leaning over. Another day passed.

The next day we picked up the remainder of the boards, dumping them, as before, on the shoulder of the road, since there wasn't any room on the steep lot. It took three trips to get it all, each one seeming endless as the truck whined along like a snail, holding up the heavy city traffic, threatening continually to stop and cause a traffic jam. I was exhausted by the last load just from inching the truck along and worrying that we wouldn't make it. We returned to pulling nails, a job which I could now see was far worse than Mike had allowed.

"When I was little, workin' for my dad," he said, "I used to clean two-thousand feet of board a day."

I could see that if he had, he must have been a glutton for punishment. I could also see that these boards were no bargain, if the nail-pulling time and the injuries from grabbing unseen nails were included in the cost. Luckily, my steel-soled shoes kept the nails I continually walked upon from piercing my feet.

I had to get to work on the sewer line and the house itself. We pulled a few more nails, using the little-traveled street as a workbench. The pile lay there on the shoulder, a guardrail styled from rubbish, in contrast to the natural setting. I had no intention of leaving this eyesore in place for long, but I didn't have anywhere else to stack it on the steeply sloping lot.

"Are you tearing down a house up there?" asked one passerby. That comment touched me in a very sore spot. Here I was working as hard as I could for the past two months, and still a person could walk past and not recognize that a house was being built. The whole project was still all in my mind—perhaps just a crazy fantasy that couldn't be done. I went to work on the sewer for a few weeks, waiting for the concrete men, digging each day in fits of masochism until I was exhausted. I didn't get the sewer done. I had to abandon it temporarily as the time came to prepare the excavations for the concrete men.

By then I was beginning to assume the attitude that no matter what lay ahead, everything that could go wrong *would* go wrong; and there was no use trying to hurry things along.

Building the Foundation 6

I adopted the principle that if anything could go wrong, it would. Digging the foundation bore this concept out. It was supposed to take a day or two, but took another three weeks of hardest work yet. Tree roots fought me. Rental tools broke daily, and their owners weren't helpful. The rock layer that the soil test had said I should reach at 4 feet was 8 to 12 agonizing feet deep.

FINDING TOOLS

The usual tools, shovels, and picks wouldn't work down inside the long skinny holes; I spent days just searching for the tools I needed. Finally, in desperation, I contacted the electric company, since I reasoned that they had to have such implements to dig the holes for their power poles. I trudged out to their tool warehouse—to me it was like a toyland full of tools. There was every tool imaginable that I could need. The tool supervisor was enchanted with my project.

"That must be some foundation," he said, when I told him I needed a shovel with a 12-foot handle. "My son-in-law built a house like that, on stilts, a few years back—of course, he had someone else do it for him. You're doing all the work yourself?" I nodded. "Well, well...that sure takes a lot of courage." He showed me the tools I needed.

"Thanks," I said, sighing with relief. "With these I ought to get the job done." He gave me a form to sign and said I could use the tools as long as I needed them.

"Don't hurt yourself, now," he said, as I was leaving. He was a wonderful relief from the rental companies' sharp businessmen.

A few weeks and a hundred boring details later, I poured the concrete. During that time the plumbing inspector came by. In contrast to the aggravating bureaucrat in city hall, though, he was friendly, a real human being. But he didn't even inspect the french drain.

"It looks like you're not in my district," he said. "Another inspector has this area; he'll have a look at it, and he will be the one you talk to from now on. Building this yourself, huh?" we talked about the house and my plans; he was so pleasant I wished he were the inspector.

"Couldn't you just look at it and tell the other inspector your opinion?" I asked.

"Oh, you can cover it up," he said. Apparently, it didn't matter either way. "He won't care." A sinking feeling caught me; all the energy, days of picking in the cement-hard clay, the sweat of hauling gravel to fill the ditch—all had been for nothing. The price for this bit of experience was too high.

During the three weeks spent digging the foundation, I decided that if I could complete the 15 one-foot-diameter holes, each over 10 feet deep in rock-hard clay, as well as completing the drainage field, I could certainly pour the concrete myself—I could do *anything* myself. This was my first great test.

The concept, at least, of a foundation was simple enough. Dig some holes, build some wooden form-molds (Fig. 6-1), and add concrete, taking care to locate everything according to the blueprint. And, technically, that was indeed all there was to it.

But I worried continuously, and lost much sleep those last few days before the "big pour."

The real problem was my own anxiety. Concrete was, perhaps, a symbol of iron-clad commitment. Mistakes with concrete are, to say the least, hard to repair, especially on the steep hillside where removing unwanted pieces by hand would

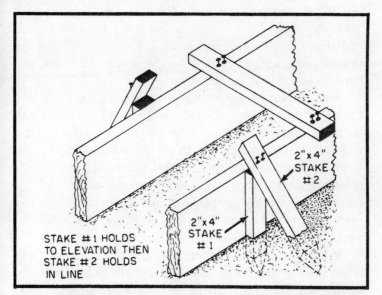

Fig. 6-1. A typical wooden form into which concrete is poured to form the foundation. These forms are removed once the concrete has set.

be impossible. And in spite of my rational knowledge that there shouldn't be any problems, many people had advised against doing my own concrete work.

"You're doing even the *concrete*? Boy, I don't know. Maybe you should have an expert do that. You have to have a solid foundation, you know. Any mistakes with it will plague you all through the house."

Also, the thought of me coordinating two ready-mix cement trucks, a concrete pump (the long distance from road to house required a pump), and two other experienced workmen was terrifying, even though I spoke with them ahead of time and they assured me not to worry. I slept little and worried constantly as the day of the "big pour" arrived.

But the building inspector who came to look at the forms the day before the big event was actually impressed by the formwork and excavation.

"This oughta be the most secure house on the hill," he said, filling out the necessary papers.

Evidently, operating "by the book" is unusual in the construction business, since the man who operated the concrete pump was startled when he looked at the holes.

"What possessed you to dig them so deep?" he gasped. "Nobody ever makes a foundation like that." I felt both proud and stupid for the excellent work and overdevotion to the engineer's requests. Oh, well, I thought, at least I'll never have to worry about the house sliding away.

With the pump hoses strung down to the waiting forms, the concrete trucks arrived. The two workmen and I wrestled with the hoses, filling the holes one after another, replacing in a few minutes the dirt removed with so much effort, filling the formwork past the delicate 18-inch mark—without even a creaking sound from the secure wood—"puddling" the concrete to remove air bubbles, and adding the last few reinforcing steel bars and metal brackets for the wood that would "cap" the foundation.

Within two hours, I was smoothing it out, the trucks and helpers were gone, and the wet cement was ready for handprints! While I was paying the truckdriver, he said, "Say, why don't you guys open an account with us?"—as if I were an expert.

I went home and washed off my shoes, filthy from the months of mud and the concrete splashes; the dirty work was finished. The *foundation*, that proverbially most important part of the house, was done; the house could now really move ahead.

THE FOUNDATION

Before beginning with the plumbing, heating, and electrical systems, there is one important aspect to building a house that cannot be overlooked—the *foundation*. With walls, floors, or roofs, you can take apart a finished house and examine every part that goes toward making it. You can make your own duplicate by simply copying the structure you see.

To make a floor, for example, you can put a bunch of edge boards (joists) together as in Fig. 6-2, which shows a house floor with the house removed. All the parts are in plain view and easy to copy. But if you tried to build a foundation from the parts you see examining a house, you'd find most of the parts are either hidden, not obvious, or simply missing.

A foundation usually looks like the rendition of Fig. 6-2. The floor joists bridge from one wall to another, supported in

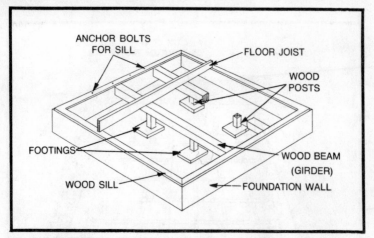

ANCHOR BOLTS FOR SILL

FLOOR JOIST

WOOD POSTS

FOOTINGS

WOOD BEAM (GIRDER)

WOOD SILL

FOUNDATION WALL

Fig. 6-2. With the foundation walls set, sills are fastened down with anchor bolts to "cap" the foundation. Wooden beams, sitting in concrete footings, hold the girders. The floor joists bridge the house, resting perpendicularly across the girders.

their centers by large beams held up by posts. The posts rest in patches of concrete called *footings*. The posts are usually 18 to 24 inches long, with the concrete foundation wall being the same height. Most houses don't look like they are as far off the ground as shown here; usually a hole is dug several feet deep to set the entire foundation into. The top of the foundation walls remain at ground level (plus six inches as most codes require).

If you went underneath the house and made a survey of the foundation parts, you'd find: *concrete walls*, each with a board bolted to its top; individual *concrete patches* (*footings*); and *wooden posts* and the *beams* (*girders*) that these posts support.

The space underneath a house, within the foundation, is called, logically enough, a *crawl space*. Down there, you'd probably see some other things not related to the foundation—plumbing, heating pipes, and the underside of the floor joists. These four basic parts (concrete walls, footings, wooden posts, and beams) are all you'd see of the foundation itself; they make up the finished unit. It takes a lot of work to get them in place, however—more than any other major part of the house; what you see is only a small fraction of the total work involved.

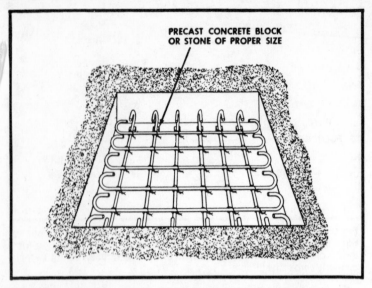

Fig. 6-3. Sometimes steel bars or rods are placed in forms before the concrete is poured. This adds additional strength to the foundation or footing, especially needed in tall buildings or hillside foundations.

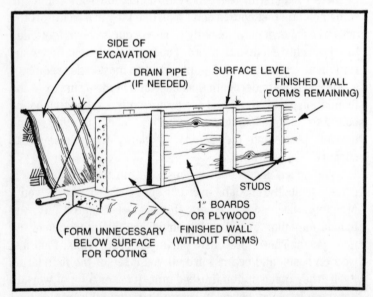

Fig. 6-4. Walls are formed when concrete is poured into wooden molds. The molds are removed when the concrete has set. In this illustration, the wooden form has been partially removed, leaving part of the newly formed wall exposed. If a drain is needed, it is placed before the wooden forms are built to accept the concrete.

CONCRETE

Normally, the concrete you see is solid all the way through—no studs, no skeleton, just solid concrete. Sometimes steel bars (Fig. 6-3) are added to increase the strength, but only in special cases where the concrete needs a particularly strong backbone, as in a tall building or a hillside foundation.

Unlike wood or other materials, the concrete never comes in smaller pieces; no one brings each wall or footing in one piece and sets it there the way it is. Instead, it comes in liquid form.

Both the wall and footings under any house are originally molded into shape by pouring liquid concrete into special wooden forms (Fig. 6-4). The walls are poured into wall forms (Fig. 6-5) and the footings are poured into footing forms (Fig. 6-6). When the concrete is dry, the forms are removed. If the forms were left in place, the foundation walls and footings would have a layer of wood on each exposed side.

Walls and Footings

The walls and footings extend several feet into the ground; a ditch is dug for the walls to rest in, and holes are made for

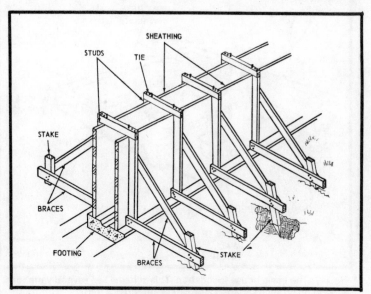

Fig. 6-5. A typical wall form.

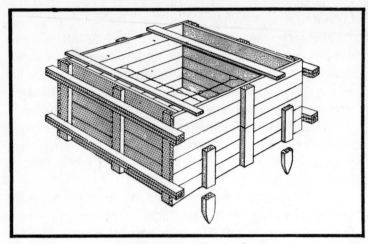

Fig. 6-6. A typical footing form.

the footings. The wood forms may not go below the ground level, though; if the earth is solid enough and the hole or ditch carefully dug to the correct size (to minimize the concrete needed), the hole itself can act as the form. Here resides a lot of unseen time and effort.

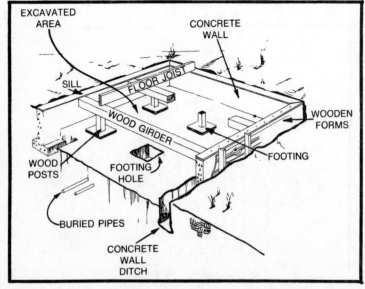

Fig. 6-7. The basic house foundation. This scheme has several variations, depending on such variables as location, preference, etc.

Sometimes, a drainage pipe is buried around the outside of the walls to protect the soil that the concrete rests on from erosion. Any underground utilities—sewer or water pipes, gas or electric lines—cross below the walls.

So altogether the house foundation includes: the excavation for the whole assembly; the buried sewer, water, and utility pipes; a maze of metal running underneath everything else; the drain pipe buried just outside the excavation to protect it from water erosion; and the foundation wall. Unseen items include the forms the walls were poured into and the ditch they sit in, the footings and the posts they support, the beams (girders) that the posts support, plus the forms that the footings were poured into and the holes they sit in.

Figure 6-7 shows a basic house foundation. Like the roof, this basic scheme has several variations, one or another describing most houses around today.

If the excavation is made extra deep, 8 or 10 feet, a basement can be made by lengthening the supporting posts, increasing the depth of the foundation wall, and paving the floor of the excavation. The house remains at the surrounding ground level (see Fig. 6-8).

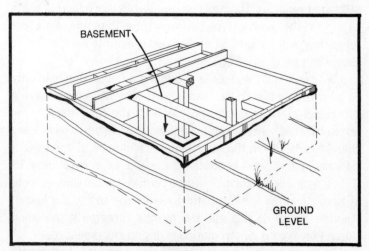

Fig. 6-8. A basement can be included in your housebuilding plans by lengthening the support posts, increasing the depth of the foundation wall, and paving the surface of the excavation.

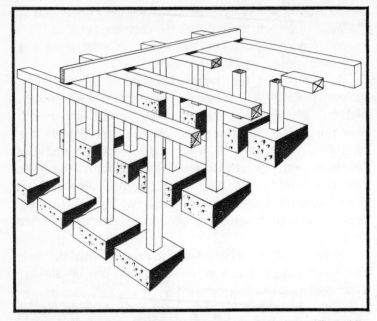

Fig. 6-9. On a hillside foundation, longer posts are needed, as well as additional footing posts to support the beams. Cross-bracing will be needed to prevent swaying if the posts are more than 5 – 6 feet long.

A house can be built on a hillside by using longer posts as stilts and replacing the concrete walls with more footing posts to support the beams. If the posts are more than five or six feet long, they will have to be cross-braced against each other so they won't sway (see Fig. 6-9).

Why go to all the trouble of digging a foundation hole only to later bridge it with the floor joists? Why not simply pave a nice level piece of earth and use it as the floor itself? One answer is that houses aren't supposed to have any wood closer than 18 inches from the soil; termites will walk in and have a banquet if the wood is touching the soil or is even near it. That's why wooden forms must be removed when the concrete is hard, to eliminate this pathway from the soil to the house. Another reason is that rot and mildew threaten if the wood doesn't have good air circulation to dry up any moisture.

In areas where coal is used to fuel furnaces, basements are needed to store the coal, as well as to house the large furnaces. Elsewhere, the space below the house provides a place to conceal the furnace ducts, the plumbing pipes, and the electric wiring.

Concrete Slabs

Some people avoid the dig-a-hole-and-bridge-it method by using a *concrete slab* foundation (earth covered with a slab of concrete) as shown in Fig. 6-10. This method has several drawbacks, however. Plumbing pipes must be placed before the concrete is poured—mistakes are hard to fix. If a smooth floor is desired, you'll have to pay an expert to finish the wet concrete. If you're on a hillside, leveling a patch of ground is far from practical. The floor tends to be cold, since it's in direct contact with the earth. And there isn't any handy space under the house to hide furnace pipes, wires, and other things. Its main *advantage* is that little digging is required and those difficult-to-make forms are unnecessary.

Aside from these comparative design considerations, the basic functions of a foundation are to hold the house up and to support the floor joists solidly so nothing sags. In the case of a slab foundation, the basic function is to support the walls

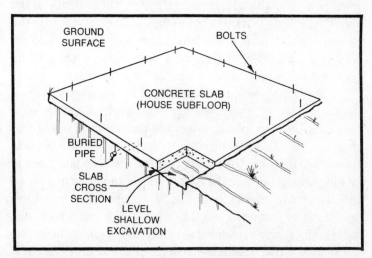

Fig. 6-10. A typical concrete slab foundation. The place where the house will eventually sit is covered with a large piece of concrete. Although this method has its advantages (e.g., little digging and no forms), there are many drawbacks, such as lack of a crawl space, the need to place plumbing pipes beneath the concrete slab, and a particularly cold basement floor. Bolts are set in the wet concrete to cap the foundation.

directly without sagging. To this effect, the footings give the posts that rest on them a broad, flat area to rest on. Without these footings, the narrow wooden posts would sink into the soil under the weight of the house. The foundation walls act similarly, as *elongated* footings, to transmit the weight of the house to the ground without sinking.

Making Footings

There is more to making footings than simply laying down a patch of concrete. The footings must be designed so they won't sink into the ground under the weight of the house. The person who designs the house decides how many footings are needed, taking into account the amount of weight each footing can support firmly, along with the weight of the house. Unless the soil is especially soft, or on a hillside (where sliding must be considered), most footings are two or three feet square and go only deep enough into the earth to rest below the frost line, where the expansion and contraction caused by freezing is eliminated.

In the rare instance where the soil is very soft, the footings must be wider, with a larger surface area. On a hillside, if the soil is ever likely to move, the footings must be attached to the underlying rock layer (*bedrock*), which may be very deep (10 feet in my case). With rock as a base, though, the sinking problem disappears, since rock is hard, hard, hard.

If the footing must rest on rock but the rock layer is quite deep, 10 feet or more, *pilings* are used. A piling is a telephone-pole log, pounded into the ground by a monstrous machine called a *pile driver*. It is pounded in—no hole is made first—until it strikes the underlying bedrock. The piling has to be long enough to reach the rock layer *plus* a little extra above ground level to support the house. Skyscrapers rest on pilings due to their immense weight, which only solid rock can withstand. House foundations rarely use pilings, however, because they cost a fortune—sometimes 10–15 times as much as ordinary foundations (see Fig. 6-11).

When the ground freezes during the winter, only the top few feet of soil will freeze; the *frost line* is the maximum depth to which the ground freezes. It varies with climate, being deeper in snow climates than in temperate regions.

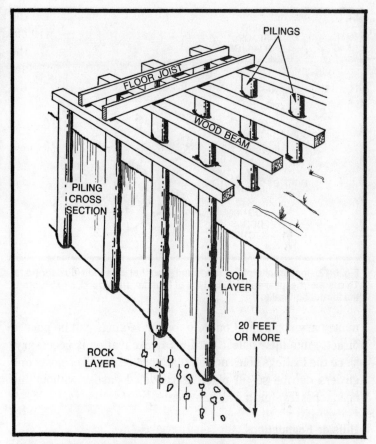

Fig. 6-11. A piling foundation is used for houses that sit on a hillside or on soft soil, as well as for skyscrapers. The piles are driven into the ground by a pile driver, until they hit bedrock.

As far as holding up the house is concerned, the foundation walls could be replaced with individual footings holding up a beam situated where the top of the wall was. The foundation walls enclose the space under the house, however, protecting whatever is concealed below; and they help to insulate the underside of the house floor from cold winds, performing a valuable function that footings wouldn't.

However, the cheapest and easiest foundation to build is one that does its basic task of holding up the house, and no more; it omits the walls, holding things up by footings, posts, and girders only, as in Fig. 6-12. The savings in both time and

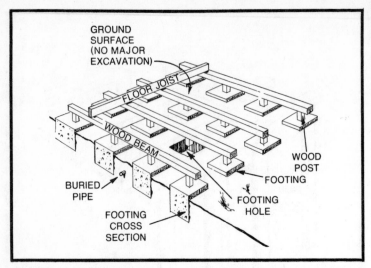

GROUND SURFACE (NO MAJOR EXCAVATION)

FLOOR JOIST

WOOD BEAM

WOOD POST

FOOTING

BURIED PIPE

FOOTING HOLE

FOOTING CROSS SECTION

Fig. 6-12. Another alternative is a foundation that uses individual footings. This is the cheapest and easiest type of foundation to build, since it omits the foundation walls.

money are apparent. Concrete for the footings can be poured directly into their holes; hardly any excavation is necessary, since the footings holes are all that need be dug. The posts and girders can be set in place and leveled easily, without the headaches leveling a wall can create.

Hillside Foundations

As a contrast to the normal 2 feet deep, 2 feet square footings (8 cubic feet) the average house rests on, I'll describe the process I used for my hillside specialty.

The foundation for my house consists of a 2-foot-high wall at the uphill side of the house, and footings around the other three sloping sides, similar to Fig. 6-9. I could have built a perimeter foundation, stepping the wall in stages down the slope, but the trench to pour the concrete into had to reach down to rock; digging a few footings to that depth was hard enough.

Hand excavation is downright primitive, but I mention it for two reasons: First, you can save money if you do it yourself. Second, as in my case, the building site is sometimes inaccessible to heavy power equipment. Only if you're a glutton for inefficient, mundane, tortuous work, or if the job is small, will you like excavating by hand. A person with a *backhoe* (a power digging machine) can dig holes in one minute that would take you hours to dig by hand. Even though such power equipment may cost $20 per hour to rent, it might only take three hours for a backhoe to duplicate two weeks of human toil.

I couldn't use earthmoving equipment because the house was 80 feet from the nearest road, and on a steep hill where no driveway could be built and no crane-like machine could reach over to it. (Such is the price of seclusion.) I dug a simple trench for the wall part of the foundation, using a pick, a shovel, and an ax. The ax was for tree roots, something I hadn't expected; but in a densely wooded area there may be more roots than soil. Where I came close to fir trees, the roots were as much as a foot wide; elsewhere they averaged 2—6 inches. Heavy equipment eats roots with ease; the average contractor wastes no time on such nonsense. I spent several days on a single root at one point, however.

Since the footings were supposed to be 5 feet deep and 1½ feet wide, I rented a motor-driven posthole digger, which looked like a giant screw with handles on each side. With one person holding onto each side for dear life, the motor started and the screw began to turn, winding into the ground, throwing dirt up out of the hole it dug. We had the feeling we were trying to keep an airplane propeller from turning; every so often it would strike a rock and the machine would hurl its controllers to the ground! After several weeks, I knew it didn't pay to do this digging myself.

One time-consuming problem was that below 5 feet the power screw failed to throw the dirt out; it had to be removed by hand.

Had it been possible to use heavy equipment, the footings couldn't have taken more than a day to dig.

Level-Site Foundations

With a level construction site, the concrete is poured out of the concrete truck right into the forms. The truck drives right up to the house. Pouring concrete is hard work, though, especially when the ready-mix truck can't reach all the forms. Then the truck discharges the concrete into the wheelbarrow, which someone wheels over to the forms, hoping not to spill it, and dumps it in place. It takes several trips with the wheelbarrow to finish the foundation, and the concrete weighs about 100 pounds per cubic foot.

Another method of moving the concrete from the truck to the forms is with a *concrete pump*. This device is a good-sized truck all by itself. It costs $40 to $50 an hour to rent. It pumps the concrete through a 4-inch rubber hose. Up to a distance of 50 feet or so, the pump truck has a crane-like boom that holds the hose up, supporting most of the weight so the worker only has to point the hose where he wants it. Hold it over the form to be filled and wait for the concrete to come pouring out.

If the distance from the truck to the forms is more than 50 feet, making the boom useless, the hose is strung over to the forms. The pump operator tends to these details. The concrete-filled hose weighs 40 or so pounds per foot. It's essential to have three strong people to wrestle with it—one to point the nozzle, one to stand right behind him and assist, and one to stand back a few feet to drag the trailing portion of the hose and keep it from kinking. A fourth person would be helpful to paddle the concrete to remove air bubbles. When a water hose kinks, the water flow stops; when a concrete hose kinks, the pump jams and you have to pay for the time it takes to get it working again; at $50 an hour, that's a wicked problem.

When using the pump, the operator will recommend the kind of concrete to get. In my case it was a special mix with rocks no larger the ¾ inch, and 3000 psi concrete. The concrete doesn't spray out, as does water from a nozzle, for instance; it surges out in firm but not overpowering globs. I was terrified of the whole process until it was done. But the pump could be shut off quickly, as in the case where we had to move 20 feet to the next form. It took two hours to pour 11 *yards* (cubic yards is the way concrete is measured: *yards* is the trade slang) plus two hours to set up the hoses. It was over before I knew it. I spent the following two hours smoothing out the surface of the wall where the wooden *sill* would sit to cap the foundation, trying to make it perfectly level. I learned later that I didn't have to worry about perfection at this point, since carpenters normally achieve it with mortar after the wall is dry. It saves time to get it level, but it isn't anything to worry about. Little nails may be placed inside the form along the level line, to indicate the point to stop filling with concrete. Put them just below the top level or not at all, since they'll have to be removed during the pour if they get in the way. Pulling nails while a $50-per-hour machine waits is indeed a waste of money.

CONCRETE VERSUS CEMENT

There is a difference between concrete and cement. *Concrete* is a combination of rock and cement; *cement* is an expensive fine powder. *Rock* is cheap; the fewer the rocks in a given portion of concrete, the harder, stronger, and more expensive it will be. Its strength is graded by the force a piece of dried concrete will withstand without being crushed. A standard grade is rated at 2500 pound per square inch, or *psi*; 3000 psi concrete is more expensive.

Ready-mix concrete is the easiest to work with and comes in a mixer truck ready to use. The cost is twice that of raw concrete you can mix yourself; but it saves much time and effort. A world of premade concrete forms and concrete accessories are on the market that can save time, and often money too: cardboard form tubes, steel brackets, and form ties are only a few of the more valuable ones. Consult the telephone directory under *concrete accessories*. If you use reinforcing steel (called rebar or rerod), be sure it doesn't touch any dirt or formwork; it will rust over the years unless it's completely encased in the concrete.

Concrete is a curious substance. It doesn't act like a normal liquid. You can pile it up, for instance: In a wall form you can pour one end higher than the other and the ends won't automatically equalize, as would water. You can pour it on a hillside and it won't run down like water, even on a 30° slope. A common wall-and-footing combination form can be poured all at once, while water would run out the bottom.

Concrete dries slowly, staying "liquid" for at least 30 minutes, and not hardening for several hours, depending on

Fig. 6-13. One important step in pouring concrete for a foundation is to avoid air bubbles that can be trapped when the concrete sets, creating weak spots. These can be minimized with a mechanical vibrator or a paddle plunged in and out of the mix as it is poured.

the temperature. Because it is so thick, air bubbles can be trapped, hardening in place and creating weak spots. These can be minimized with a mechanical vibrator or a stick-paddle plunged in and out of the wet mixture as it's poured (Fig. 6-13). Plunging also helps keep the gravel evenly distributed, further increasing strength. Concrete hardens to testing strength within 28 days; but house construction can begin in two or three days, when the forms are removed.

Mistakes

Unlike wood and other construction materials, concrete is a bad thing to make a mistake with; I was afraid of working with it until I'd worked with it once and everything turned out okay. With wood, mistakes are fixed by pulling a nail or making a cut; with concrete, using a jackhammer or an old-fashioned sledgehammer is the only way to make changes. Moving the broken chunks of mistakes, especially up a hill, and carting them off would be backbreaking, if not impossible. But as long as forms are checked to insure the correct location and level, the concrete has to go in the right place.

Several of my footings came out two inches from their required location; luckily, they were wide enough so that even with the wooden posts two inches off center everything was fine. Had a wall been built too low or out of level, mortar or layers of wood could have been added to correct the error.

I was afraid the concrete would run out between the forms, run down the hill, dry so fast I wouldn't have time to even roughly level it, take so long to pour that I'd have to pay the truckdriver for extra time (eight minutes per yard is free, after that you pay), that the forms would burst in some undreamed of way, or that I had measured wrong somewhere.

It helped a lot to see everything done. The best way to rid yourself of fear, other than believing whole-heartedly in this book, is to follow a concrete truck to its destination and watch the process. The trouble is that if you follow this advice with every process in the house it may take longer to learn than it takes to build the house. I just went ahead, assuming that the worst that could happen would be that the trucks would arrive along with the pumping machine, and I would chicken out, forfeiting the costs.

CONTRACTORS

Having a contractor do this work is an attractive alternative. The cost can vary dramatically from one contractor to another, however. A contractor who has a small amount of business *might* work for no profit, just to keep his workmen employed until he gets more work. A single house provides little work, actually. A person who is busy or greedy may set a ridiculously high price, since he doesn't need any additional work and can afford to lose your business. Some figure that anyone building a house single-handedly is an easy mark, and will expect you to accept a high bid through your ignorance if you fail to compare. In regions where construction work is seasonal, the chances of getting a good price in the slack season are good, since keeping workmen working is important.

Carpentry & Framing

7

I still hadn't figured out how to finance the whole project; but I was pretty disenchanted with used lumber and all those nails. I ordered the first batch of new lumber from a lumberyard. After weeks of hard, hard work, I walked into the lumberyard, this time with confidence; those weathered men around the counter failed to scare me any more. I opened an account; I was called by my first name, one of the boys.

Nevertheless, I was thankfully absent when the lumber was delivered; I didn't know the slightest thing about unloading it, or storing it either. Nor was there any good place to put it, since the shoulder of the road was piled with used boards. My presence would have, at best, only allowed me to watch helplessly as they dropped it off the truck.

"Get out of the way, kid," I could hear the driver saying as I stood near the truck about to be run over. "Where's your boss, anyway?" I wouldn't know what to do with myself and, as with the cement truck, I couldn't stand the thought of being looked down on now; I had a house to build and needed all the confidence I could get.

It was a shock that next morning to see the several stacks of lumber waiting for me to turn them into a house—as if I could do such a thing! The previous night I'd gone to buy tools: a 20-ounce all-steel claw hammer, a framing square, a level, two sawhorses, and, most importantly, a carpenter's leather tool and nail pouch. Without the pouch I would have been lost; but I knew I was a carpenter with it. I strapped it on and took a deep breath—the house began!

I was unsure of myself at first, though. It took a week to build the floor, a task I could finish now in several hours. I worked with book in hand, at first, measuring everything several times, bending more nails than I drove, and using the wrong size nails at first and having to redo everything with the correct size.

One night I called my friend the house framer; maybe he could show me some tricks of the trade and speed things up a little; I was progressing but remained uncertain. Big Dee, as he was known in high school, was excited when I told him what I was doing, and gladly offered his aid. Unfortunately, he was going to Europe in a few weeks; but he said he would come over after work one day and help me put the first wall together.

"What's with all that used lumber?" he called out, as he started down the path from the road to the house. "Are you planning to use that?" He was wearing his carpenter's pouch, ready for business.

"Yeh, I was going to use it for wall sheathing," I replied.

"Hmmmmm... I guess that's all right. Get it free or something?" By now I was beginning to realize that it was foolish to buy that used lumber.

"No, I didn't." My reply was grim, embarrassed.

"Well, I guess it's okay. The only thing is, I know some guys doing their own houses, and they've had trouble getting money from the bank. I don't know how a bank would like this old stuff. But it's your house! How're ya doin'?"

At this point an irrational fear of used material sprang up inside me that was to keep me from using much more of it.

"Oh, fine. Glad you're here, Big Dee," I said.

"Okay, let's go," he said. "Get four long two-by-fours." We were off. He marked where the walls would sit on the floor with a chalkline, cut the edge of the floor to match, and started to mark the wall pieces for doors, windows, and partitions, all in less than 10 minutes. His working pace was feverish, contrasting my relaxed one of the past week. "Okay, have you got a plan?" I handed it to him. He put his ruler to it, measuring the distance to the first window from the end of the wall. "Okay, so it's two feet to the edge of the window."

"Well, no it isn't," I said. "I've changed the plan and it's, ah, let's see..." I figured for a moment. The windows had changed again since the plan was redrawn, and I had to make sure everything would fit correctly. "It's eighteen inches."

"Okay, eighteen inches," he said, and marked the lumber accordingly for the stud position. He went along the whole wall, similarly, taking another 10 minutes total, including my refiguring for windows and things. "That's it," he said at last. "That's all there is. Just a few tricks that save time."

The next day (Saturday) Diane and I built the wall he had laid out and applied the plywood sheathing. Nailing on the plywood may sound easy, but it's very slow; the codes require that a lot of nails be driven to hold the plywood and wall framing solidly together—one nail every 6 inches around the edge of each sheet.

Using that wall for a model, we made the facing wall across the floor. Big Dee returned on Monday to show me how to lift these two into an upright position, since each wall was made lying down.

"Beautiful," he said. "They look like a pro did 'em—I guess you had a good teacher. Ha, ha."

I rented some *wall jacks* as he advised, not knowing how they worked. He inserted a long two-by-four in the right spot, and we jacked the walls into place one at a time. We took great care not to let them fall over, off the platform. As the walls approached their upright position, where another single jerk with the jack might send them over, I attached some two-by-fours to every other window or door opening. Being careful to stand back far enough so that, if the jacks slipped, the falling wall wouldn't land on me, I held the center brace until the middle section was almost vertical; then I nailed my end to the floor. The two ends were brought up in a similar way—very gingerly. Finally, we made the wall exactly vertical by disconnecting each brace in turn and moving the top of the wall until it was perfectly aligned.

"Well, good luck!" he said when that was done, in about half an hour total. "I'm off for Europe tomorrow." And he left.

In another two days, I had the other two shorter walls completed and in place. These few days were some of the best of all. I was cutting boards for door and window openings, corners, and other wall parts—and doing it right! I became excited each time I cut a board that actually fit where it was supposed to. The weather was still pleasant. It was amazing how morning arrived at the house with one wall standing, and that evening it had two—and I did it. I had somewhere gained a magical competence since building that tree house.

With the four walls up, I thought I was almost done. I had spent more than two months digging and rooting around in the earth below with no visible results, getting more and more depressed. But now, in a single five-day period, I had gone from a rudimentary platform to a house-like thing, complete with four walls and some window and door openings. If you stood in the right place, you could be shielded from the wind. All I had left to do was the roof, I thought.

I worked faster and faster with practice. Things were going well at last. In fact, the stacks of lumber, two weeks before a worrisome challenge, were now consumed. I needed more lumber now, heavier beams for the roof and ceiling. Maybe Wrecker Mike would have something I could use.

When I tried to contact him at the old place, though, the wreckage he'd been overseeing had vanished, and he with it. I found him, through the wrecking company, at a new site, commanding a new pile of debris. He had only three timbers I could use, but I strapped them onto my truck to carry them back to the house.

I drove through the early morning rush hour traffic. Each of the three 4- by-12-inch beams were longer than the truck, extending several feet over the front. I thought they were secured to the top of the truck, but I drove as gingerly as I could anyway, just to be safe. My nerves were on edge with this load and, at one intersection, the light changed from green to yellow as I approached. I slammed on the brakes. My worst fear came to life! The beams gracefully glided into the intersection ahead of me, stopping traffic in all directions. It had taken five people to load them onto the truck; so I knew I would never get them back alone. Horns started honking. I had visions of being towed off to the police station, and having to pay to have the beams towed away as well. I walked out to the beams and tried to hoist one up—but I could barely budge it. I looked around, seeing the traffic jam worse. Then, out of nowhere, three burly workmen appeared. The grabbed the boards and replaced them on the truck. Whew!

I went on my way, more slowly and gingerly than before, arriving, finally, at the house. I ordered the rest of the beams from the lumberyard, and went down to dig at the sewer to calm my nerves for a day or two until the order arrived.

"Will you please be sure to deliver the lumber on a crane truck," I said. "My house is seventy feet from the road, and I have to carry the stuff down on my back if you don't let it down with the crane."

"Sure, no problem," came the reply. "The truck isn't free for a couple of days, though. Can you wait for it?" I said I could.

But the lumber arrived on a standard truck. And they dumped it at the road. It took me a whole day just to carry the large, 20-foot 4-by-8 beams down. It was needless hard work; I vowed never to order lumber there again, even though it was a friendly place in other ways.

Three days after I carried the lumber to the house, the second-floor beams were in place and painted; another week and the floorboards, which also formed the first-floor ceiling, were painted on the ceiling side and nailed down. Daily progress was less than with the lower walls, but I felt pretty good about what did get done.

I often came across a situation where I tried to make a piece of unworkable wood fit where it simply wouldn't. I had either made an error in cutting or the piece of lumber had too many knots and there weren't any places to nail through. Pounding it with a hammer with all my might helped relieve my frustrations, sometimes even squeezed the piece into place; but it usually destroyed the piece and the surrounding material with it—depending on the extent of my stubbornness.

Finally, I would give in to the demands of the situation and start over, abandoning my childhood dreams of having wishes perform magic, realizing that my wishes had destroyed several days of work instead. Other real-world frustrations included delivery trucks that failed to arrive with needed materials, winds that upset leveling operations by making the house parts sway, and cold weather that numbed my hands so that they no longer obeyed my commands. Striking back at Mother Nature, as I tried to do with the hammer, was futile, useless, and senseless in the face of her indifference. Letting out bad feelings by screaming, hammering, kicking, or running around was absolutely necessary. Frustrations came too often; it would have been pointless to try holding them inside—even dangerous. When I turned the rage upon myself, I would find myself wanting to run off the platform and end it all. More safely, I destroyed a few boards in fury.

Eventually, I became more relaxed. I accepted as a fact of life, a painful one, that the physical world didn't care a fig for me. As a human being, I hadn't been invited to this world; nor had I been promised happiness or satisfaction. I accepted calmly that things would normally go against my will. Problems would arise as fast as I could solve them; nothing would go my way just because I wanted it to. At least I knew that for everything that could go wrong, there would be a way to fix it; no matter how much trouble I encountered, there would be a next step small enough to tackle. I might be stuck forever on a treadmill—but that was the risk.

The second-floor walls took longer than the first-floor walls, since none of them were square and every stud had to be cut to fit. Diane nailed most of the plywood on when she had time.

My original house fantasies included hordes of friends dropping by each day to help. But reality was less festive. Everybody had a daytime job; no one else had a commitment to my house. I couldn't expect them to use parts of weekends, their only leisure time, to work with me. Building a house was no weekend project; anyone could come three weekends in a row and not see much progress. In addition, I planned most of the activity for myself, assuming I would be alone, leaving little spare work for casual, inexperienced visitors. And I hadn't time to play instructor.

An exception was some new friends who'd recently arrived in town. Jeff came over several times to help; ironically, I was out on errands the first three times he came. Crossing paths at last, we installed an upstairs wall in a single day. Working with another person brought a dramatic speedup in the work, but he had to leave town for a while. I was alone again, grinding out the walls, hammering, nailing, sawing, and measuring.

The first four walls had been a joy to build; the first-floor ceiling (second-story floor) had been a thrilling new achievement. But the second-floor walls became a grinding repetition of the previous work; and I was thankful when the roof began, since it meant a major new milestone in the building process.

The roof beams went up quickly, though with much grunting and groaning. Before the beams were up, the house was basically an ugly square box; now, with just these boards, the airspace over the second floor acquired a pleasant shape that gave the house its first touch of grace; excited, I barely felt the weights I lifted!

There were several beams too heavy for one person to handle; Jack came by to help put them up. Planning for these "big lifts," we expected each one to require a lot of strength, but didn't realize how difficult it would actually be to perform safely. The most troublesome beam, a 28-foot, 4-by-12 (one of the "gliders" from the truck incident) weighing well over 300 pounds, took more than an hour to secure in place. We had to climb a 12-foot ladder to position it. But climbing wiggly ladders with one end of a 300-pounder on your shoulder is difficult. We jockeyed for position, our hearts skipping now and then as we slipped; one false move and the heavy weight would crush us both. Very, very carefully, we inched our way upwards until, at last, with one final ever-so-delicate shove, straining with all our might for control, we touched it lightly into place. Supported on edge by a pole at each end, it resembled a tightrope more than a roof-ridge beam up so high.

With the ordeal over, another friend came by and showed me a nice trick—nailing a 2-by-4 to the beam so that a third person could support some of the weight from below by holding up the nailed-on post. But as with many hints, it came too late to be of help.

The boards for the roof deck remained to be installed. It took several days to carry them from the road, and several more to paint them. (One side is the ceiling for the second floor.) I spent the first big snowstorm of the year, which lasted two weeks, putting them on. Big Dee's parting comment came back to me as I labored in the blizzard, brushing snow drifts from the boards as I nailed them on

"That's a mighty steep roof; don't fall off it now!" he had warned.

Oh, well, at least with the temperature below freezing, I couldn't get wet, although I did have to keep knocking the icicles off my mustache and eyebrows.

I was up there nailing, as the snow came down, for quite a while. Near the end of the ordeal, I heard a voice saying "Hello? Hello?" Being at least 40 feet off the ground on the downhill side of the house, I had a commanding view of the whole area and the valley below. Thinking maybe I'd see someone stuck in a snowbank with their faithful St. Bernard going for help, I tried to locate the voice. Below me, at the house next door, stood a young woman looking up in my direction. I waved.

"Hello!" I said. On the roof, I began to feel like an Eskimo, far away from the civilized world, in a frozen heartless land.

"Hello!" she said. "Would you like some hot coffee?" Was this a mirage?

"Wow, that's really nice," I said. "You don't want to bring it up here though, do you?" I didn't feel like going down. She said she would. She soon returned and I met a neighbor.

A combination of empathy and curiosity about what could make anyone work in a blizzard brought her out. I thankfully accepted the coffee and cookies in this unlikely picnic spot. She had been hearing me hammering ever since I started, and had been telling her friends that there "seemed to be some nut" building a house single-handedly next door.

"Well, I thought the house would only take two months when I began," I said. "Now that it's been four months and I'm still not finished, I can't stand going home when I know that no work is getting done. I'm the only one that does the work, and I've got to get it finished or I'll go mad."

I finished the roof decking from inside the house. Carpentry, after all, is what a house is made of, I reasoned; and the hammer-and-saw business seemed nearly completed. Except for the doors and windows, all I needed was a layer of shingles on the roof to stop the leaks. The house would soon be done.

CARPENTRY

That first day of carpentry, when I was faced with an empty lot and a pile of lumber, was terrifying. I didn't have a penciled-in house floating up above, beckoning me to fill in the spaces. I couldn't really believe the first few boards I picked up were part of the floor; they were simply boards, single lengths of wood.

I could only see the house if I removed myself to my imagination world and thought of what the book said about how these boards were floor parts. In the world where I had to start hammering nails and placing boards, there wasn't any house at all—only bare land and a pile of lumber, and a few boards nailed together looking like a treehouse.

Even later, when most of the structure was in place, I kept seeing each task as nothing more than itself, not as an important house part on which the rest of the house depended. I couldn't see their permanence; the board I nailed in place *now* could be removed *now* as well, and I couldn't visualize the board as part of the whole house, buried under the successive layers of the building process.

FRAMING

I sit here in my living room, a single, simple room with four paneled walls, a high wood ceiling, several windows, and a fireplace. The floor is carpeted, although it could be linoleum, hardwood, tile, or some other material in another house.

The carpet is a kind of indoor grass to me; it seems as though I could pull away a patch of it and reach the dirt below, as I would outside. But actually, wood houses sit a few feet off the ground; if I did dig into the carpet, I'd come up with wood—the *subfloor* that this *finished* floor is built on. If I tore up the subfloor, a layer of plywood, I'd find the skeleton of the floor itself—but still no dirt.

The *framing* layer, the part that holds things up, is just a lot of boards standing on their edge (see Fig. 7-1), each one

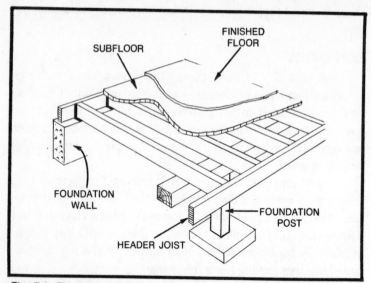

Fig. 7-1. The basic framing that supports the floor is actually several boards placed on edge that will later be covered by a subfloor (usually plywood or particle board) and a finished floor (carpet, tile, hardwood, etc).

bridging between the supports at both ends. It might seem strange to build using boards *on edge*; Don't they tend to tip over, as would, say, a nickel if you stood it on its edge? But the bending strength of a board on edge is extremely high compared to a board laid flat (See Fig. 7-2), so it's worth the trouble to set these boards in place on their edges.

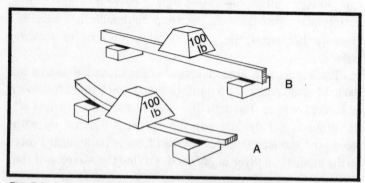

Fig. 7-2. A simple experiment demonstrates the effectiveness of placing boards on edge. With the board laid flat, spanning two blocks, a 100 lb weight will cause the board to sag. But with the board placed on its edge, spanning the same blocks, the 100 lb weight will not even flex the board.

Basically, the whole house has two main layers: a *skeleton* (framing) layer that provides the strength to hold things together; and a *decorative* (finished) layer, with little or no strength of its own, that hides the skeleton. Even the plumbing, electric wiring, and heating system works this way, with most of the working part of the system hidden.

Framing is that part of carpentry that deals with the details of how the skeleton layer is built. It's also referred to as *rough* carpentry or *roughing in*. Little of this work shows in the finished product, and thus it needn't have a nice appearance—you can beat up all the boards you want, since none will show. To meet the test of strength over time, however, the framing must be done properly, in spite of appearances.

Framing details for the floor consist of such things as:

- What size nails to use and where to put them.
- How to connect the joists securely to their supports and in an upright position.
- How to nail the subfloor over the joists.

Figure 7-3 shows the different sizes of common nails used for framing. Table 7-1 tells you what size nail to use for each chore and how to space those nails.

Floors

Building the floor is probably the simplest activity in house construction. The parts are few; you simply buy the right boards and put them in place. When I began, I assumed everything would be quite demanding and difficult, especially the floor, the base of the house itself. But it wasn't. Even taking great care, as if I were a Saturday afternoon perfectionist who didn't know how to drive nails, it only took a day to put the joists in place—32 of them, each 26 feet long—and only several more days to add the *blocking* and the subfloor (see Fig. 7-4).

I thought I'd save some time by buying these extra-long 26-foot boards. But the extra cost was wasted, since they were difficult for one person to carry. Several 10-foot boards would

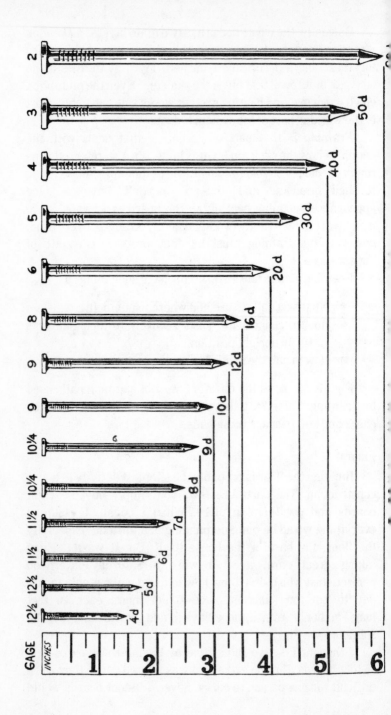

have been much easier; even a small person could carry around a 10-foot or 14-foot two-by-eight.

You may wonder how great a distance one of these joists can span without sagging. It depends on the depth of the board; a two-by-twelve spans farther than a two-by-eight. But two-by-eights are often used in houses, spanning 10–11 feet with no noticeable sagging.

Engineers and architects have tables that tell what size boards to use for spanning a certain distance. However, any plan you buy will include the proper sizes; unless you draw your own house plans, these engineering details won't be anything you need to think about.

For any house, the *spacing* between the joists can be varied in several ways, while retaining the same strength. The cheapest floor occurs when the joists are thin and the spacing is small, since thinner boards are less costly, and narrow spacing allows for cheaper subflooring. Two-by-eights spaced

Fig. 7-4. The first joists are in place and ready for the blocking and sub-floor.

Table 7-1. Recommended Schedule For Nailing the Framing and Sheathing of a Well-Constructed Woodframe House

Joining	Nailing method	Number	Size	Placement
Header to joist	End-nail	3	16d	
Joist to sill or girder	Toenail	2-3	10d or	
			8d	
Header and stringer joist to sill	Toenail		10d	16 inches on center.
Bridging to joist	Toenail each end	2	8d	
Ledger strip to beam, 2 inches thick		3	16d	At each joist.
Subfloor, boards:				
1 by 6 inches and smaller		2	8d	To each joist.
1 by 8 inches		3	8d	To each joist.
Subfloor, plywood:				
At edges			8d	6 inches on center.
At intermediate joists			8d	8 inches on center.
Subfloor (2 by 6 inches, T&G) to joist or girder	Blind-nail (casing) and face-nail.	2	16d	
Soleplate to stud, horizontal assembly	End-nail	2	16d	At each stud.
Top plate to stud	End-nail	2	16d	
Stud to soleplate	Toenail	4	8d	
Soleplate to joist or blocking	Face-nail		16d	16 inches on center.
Doubled studs	Face-nail, stagger		10d	16 inches on center.
End stud of intersecting wall to exterior wall stud	Face-nail		16d	16 inches on center.
Upper top plate to lower top plate	Face-nail		16d	16 inches on center.
Upper top plate, laps and intersections	Face-nail	2	16d	
Continous header, 2 pieces, each edge			12d	12 inches on center.
Ceiling joist to top wall plates	Toenail	3	8d	

Connection	Nailing method	Number of nails	Nail size
Ceiling joist laps at partition	Face-nail	4	16d
Rafter to top plate	Toenail	2	8d
Rafter to ceiling joist	Face-nail	5	10d
Rafter to valley or hip rafter	Toenail	3	10d
Ridge board to rafter	End-nail	3	10d
Rafter to rafter through ridge board	{ Toenail	4	8d
	{ Edge-nail	1	10d
Collar beam to rafter:			
2-inch member	Face-nail	2	12d
1-inch member	Face-nail	3	8d
1-inch diagonal let-in brace to each stud and plate (4 nails at top)		2	8d
Built-up corner studs:			
Studs to blocking	Face-nail	2	10d	Each side.
Intersecting stud to corner studs	Face-nail		16d	12 inches on center.
Built-up girders and beams, 3 or more members	Face-nail		20d	32 inches on center, each side.
Wall sheathing:				
1 by 8 inches or less, horizontal	Face-nail	2	8d	At each stud.
1 by 6 inches or greater, diagonal	Face-nail	3	8d	At each stud.
Wall sheathing, vertically applied plywood:				
3/8 inch and less thick	Face-nail		6d	{ 6-inch edge.
1/2 inch and over thick	Face-nail		8d	{ 12-inch intermediate.
Wall sheathing, vertically applied fiberboard:				
1/2 inch thick	Face-nail		1½-inch roofing nail.[1]	
25/32 inch thick	Face-nail		1¾-inch roofing nail.[1]	
Roof sheathing, boards, 4-, 6-, 8-inch width	Face-nail	2	8d	At each rafter.
Roof sheathing plywood:				
3/8 inch and less thick	Face-nail		6d	{ 6-inch edge and 12-
1/2 inch and over thick	Face-nail		8d	{ inch intermediate.

[1] 3-inch edge and 6-inch intermediate.

16 inches apart is an example—these are the smallest you can use, and they're close enough together so that a thin and inexpensive plywood (half-inch thick) can be used without sagging.

If you're in a hurry, your time is costing you a lot, as if you paid yourself union wages. It might be worthwhile to consider a way to make the spaces larger. Using larger joists, for instance, would save time by leaving fewer joists to install. You could use a single 4- by 8-inch joist every 4 feet, according to the engineering tables, and still have the same strength. Of course, the half-inch plywood subfloor would now sag over this 4-foot space between joists, so you'd have to use boards 1½ inches thick to get the same strength as before, at a cost far above that of the half-inch plywood. And since the 4- by 8-inch joists would cost more than twice the price of 2- by 8-inch joists, the time saved would cost plenty, perhaps negating savings in the cost of labor.

Walls

You should have a basic feeling for the walls now, so that the details of house construction will make more sense. A few more notes on framing may be of further help, though.

As I've already said, the 2- by 4-inch studs in the walls protect the wall from bending in the wind by virtue of their edge strength, and they support the upper floors and roof. After seeing Figs. 7-5 and 7-6, you should have a picture of these studs inside the wall. Now strip away all the wall coverings in your mind and concentrate on the basic skeleton (see Fig. 7-7). The ends of each stud are nailed to the top and bottom plates, two long horizontal two-by-fours. The roof will sit directly on the top plate; normally a second top plate is added to the first for additional strength. The studs are 16 inches apart, measured center to center.

You would think that the studs would be held in place quite firmly by these top and bottom plates, especially with diagonal blocking as shown in Fig. 7-6. But green lumber twists and warps as it dries, and the top and bottom plates are almost helpless to hold down the studs when the twisting starts. The finished wall material, the sheathing, and the wallboard on

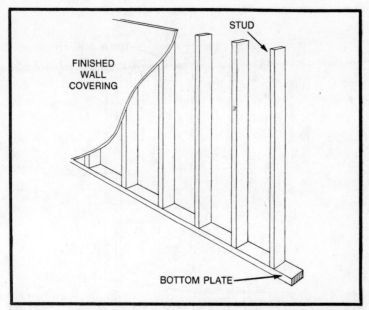

Fig. 7-5. A typical wall core. Two- by four-inch studs are spaced evenly (usually 16- or 24-inch centers) and nailed to a sole plate and a top plate (not shown). The studs are then covered with the finished wall materials.

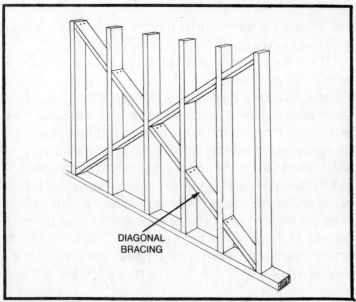

Fig. 7-6. A wall section is made stronger if it is reinforced with diagonal wall bracing.

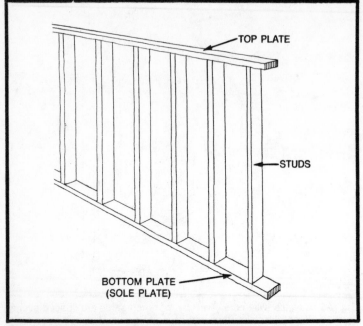

Fig. 7-7. The top plate and sole plate are nailed to the studs in a similar fashion, as shown in this basic wall frame.

both sides of the studs relieves the problem by rigidly locking each stud into position by their "sandwich" action.

If you cut the middle portion from a few studs in a row, of course, the roof would cave in there. To prevent such a disaster, a bridge is made across the opening (see Fig. 7-8). A heavy piece of wood is used to make the bridge (a *header*)—4- by 8-inch or 4- by 12-inch boards, supported at each end by a 2- by 4-inch board called a *trimmer*. The short remaining 2- by 4-inch boards above the bridge *cripples* hold up the top plate and roof. An extra stud is nailed in place to the top and bottom plates against each end of the header (with at least six nails, to keep the header from twisting out), and to each trimmer. Sometimes the header is large enough so that the top of the header reaches the bottom of the top plate, and no cripples are needed—usually with a 4- by 12-inch header and a 7-foot, 10-inch ceiling.

A door opening is converted to a window opening by nailing in a horizontal two-by-four (*sill*) at the level where the

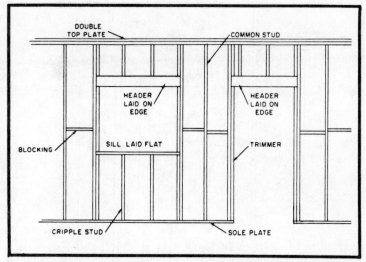

Fig. 7-8. Basic door framing. Once the wall frame (studs nailed to top and sole plates) is constructed, door and openings may be cut. Of course, these cuts would weaken the support of the basic framing, so a bridge or header is used across the opening. The sides of the opening are supported with trimmers, extra 2- by 4-inch studs. Cripples are added above the door opening and a second top plate is added for extra strength. Once the completed unit is raised into position with wall jacks, the sole plate is nailed to the floor. Then the section of the sole plate at the bottom of the opening can be cut away.

window bottom is to be. Short studs (*cripples*) are inserted between this horizontal sill and the bottom plate, supporting the sill itself (see Fig. 7-9). If the opening is more than 6 feet long, double trimmers are required. Double sills are sometimes needed, too. These openings, once built, look like someone simply cut a hole in the original studs, adding these two horizontal pieces (*header* and *sill*) at top and bottom of the opening. The doors and windows are simply slipped into these rough openings later.

In addition to the skeleton parts already mentioned, special posts are added to help fasten walls together at corners and places where inside walls meet outside walls. In corners, these are *corner* posts, in other places they are *partition* posts. Inside walls, or *partitions*, are the walls that separate one room from another within the house; the house is one large room under the roof until the area is partitioned into individual rooms (living room, bedroom, kitchen, bathroom, etc.). Some

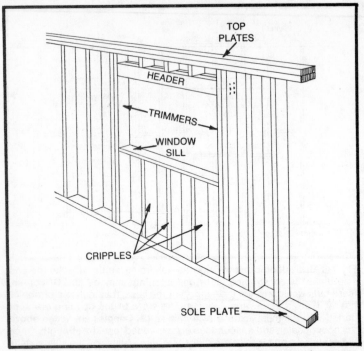

TOP PLATES

HEADER

TRIMMERS

WINDOW SILL

CRIPPLES

SOLE PLATE

Fig. 7-9. Window openings are made in the same basic way as door openings, except that cripples are added below the sill as well as above the header.

are used in holding up the ceiling (bearing walls); others aren't, as a plan will show. All partitions are built like exterior walls, except they have wallboard on both sides and no windows (Fig. 7-10).

Walls are easy to build. The studs come precut to the correct length for an 8-foot ceiling. The framing process, which took two weeks for my house, involves: cutting trimmers, headers, sills, cripples, and plates; measuring to locate windows, doors, and inside wall connections; nailing everything in place; and tilting up the finished product. The walls were built lying down. As with flooring, the only tools needed are a hammer, a level, a saw, and a chalkline.

Some builders apply the building paper and siding to the walls during the initial framing, with the walls still lying down; some even put the windows in at that time. Normally, however, builders rush to get the roof completed, so they'll

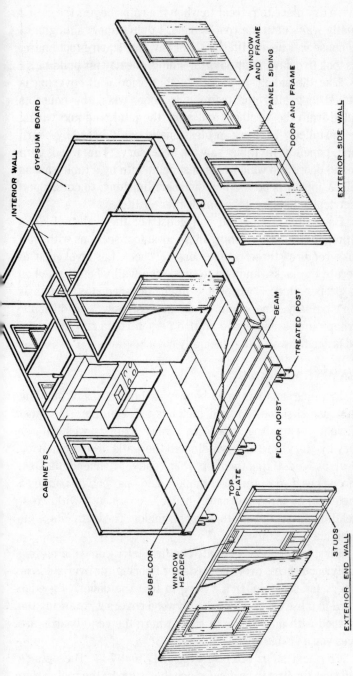

INTERIOR WALL

GYPSUM BOARD

WINDOW AND FRAME

PANEL SIDING

DOOR AND FRAME

EXTERIOR SIDE WALL

CABINETS

SUBFLOOR

WINDOW HEADER

TOP PLATE

FLOOR JOIST

TREATED POST

BEAM

STUDS

EXTERIOR END WALL

Fig. 7-10. Exploded view of a wood-frame house. A house is really one large room sectioned off into several smaller rooms.

have a dry place to retreat to when the rains begin; the siding usually waits until the roof is done. I didn't apply siding until the house was nearly finished. For my 1400-square-foot house, this took five days: four days for siding, one day for building.

Like the siding, in my case, the inside wall covering is applied long after the framing is done, when the house is sealed from the weather, and after the plumbing and wiring are installed. To work professionally with board-by-board wood paneling, a table saw is necessary. The wood trim around doors and windows is best dealt with by a table saw as well. A jigsaw is necessary for inside finishing, to cut around electrical outlets, switches, and other obstacles.

Basically, one person can construct the walls. (I did.) Tilting up the basic framing in 20- to 40-foot sections with wall jacks requires three people: one to check the level, one to operate the jacks, and one to brace the wall when it reaches the proper place. Two people might be adequate, though it won't be easy. One would be dangerous, slow, and difficult, perhaps impossible, since no one person could run the jacks and brace the wall too—especially not a beginner.

FRAMING METHODS

Framing, as mentioned before, is the process of building the house skeleton. Several methods are common, the most efficient being the platform (also called western or story-by-story) method, building all the walls in sections lying down, as I did, tilting them up when ready. My friend Big Dee, who showed me the technique, said he could frame a single-story three-bedroom house, complete with roof sheathing, in six days with only one helper. The elements of his technique are as follows.

First the subfloor is built. The first trick comes in nailing on plywood sheets over the joists for flooring (or over studs or rafters for sheathing); it's hard to find the underlying edge board in these cases, since the plywood covers it; marking the plywood with a chalkline to show where the edge boards are gives you a visible line to nail on.

The next steps refer to Figs. 7-11 and 7-12. The outside walls are the first to be done, since they support the roof. From

110

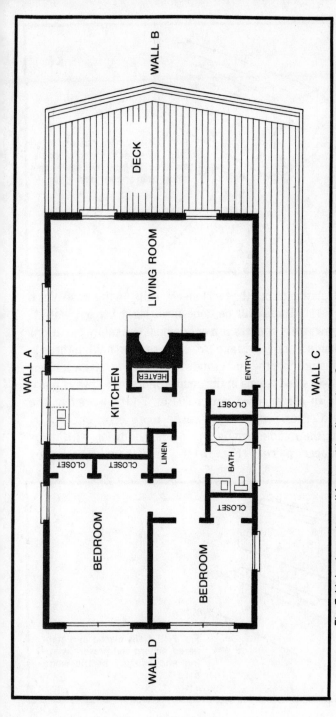

Fig. 7-11. Laying out a wall frame with a floor plan. Count the number of corners and partitions the wall has starting with the two longest walls, (in this case, walls A and C). Follow the framing procedure as explained in the text.

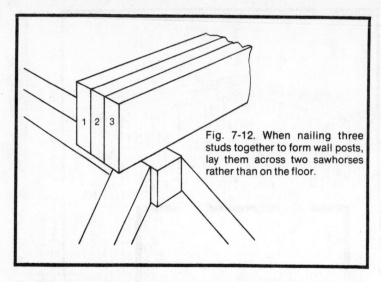

Fig. 7-12. When nailing three studs together to form wall posts, lay them across two sawhorses rather than on the floor.

the floor plan, pick out the two longest walls, in this case A and C (Fig. 7-11). These will be done first. First lay out wall A. Notice how many corners it has (2), and how many partitions (2—at points x and y.). Make two corner posts, nailing three 2-by 4-inch studs together in a sandwich (Fig. 7-12). New lumber always twists and warps in the course of drying; always place the nails in a staggered pattern (Fig. 7-13) to prevent the twisting. Corner posts can be made more economically by replacing the middle two-by-four of the sandwich with three short spacer pieces (Fig. 7-14). Lay them across two

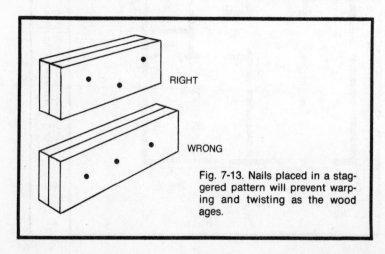

RIGHT

WRONG

Fig. 7-13. Nails placed in a staggered pattern will prevent warping and twisting as the wood ages.

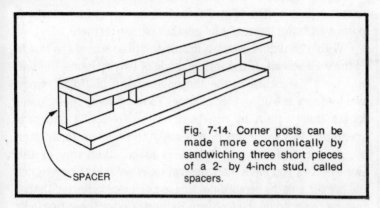

Fig. 7-14. Corner posts can be made more economically by sandwiching three short pieces of a 2- by 4-inch stud, called spacers.

SPACER

sawhorses rather than on the floor; your back will appreciate it.

Make two partition posts following the procedure illustrated in Fig. 7-15. Lay three studs against each other, thin side up, to form a corner post without nails. Add another stud on top, flush with the edge of stud 1 (Fig. 7-15A) and nail through it into stud 1 with four evenly spaced nails. Turn the pile over, so stud 4 is on the bottom. Then take stud 3 and nail it into stud 1, just like stud 4 (Fig. 7-15B). Remove stud 2 and the

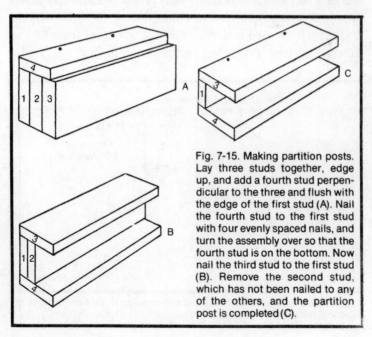

Fig. 7-15. Making partition posts. Lay three studs together, edge up, and add a fourth stud perpendicular to the three and flush with the edge of the first stud (A). Nail the fourth stud to the first stud with four evenly spaced nails, and turn the assembly over so that the fourth stud is on the bottom. Now nail the third stud to the first stud (B). Remove the second stud, which has not been nailed to any of the others, and the partition post is completed (C).

partition post is done (Fig. 7-15C). Set the corner and partition posts aside until later. (Or let another person do them.)

With a chalkline, mark a line on the floor where the *inside* of the wall will sit. If the wall is made of two-by-fours, the line will be $3\,^9/_{16}$ inches inside the edge of the platform, since two-by-fours are $3\,^9/_{16}$ inches wide. Take 2 two-by-fours, each longer than wall A by one inch. (If you're using half-inch sheathing, this extra length will make the corners flush later on, as you will see.) If wall A is longer than any of the two-by-fours you have, nail several together end to end to get the proper length; be sure not to use any pieces shorter than 4 feet, as required by the Uniform Building Code. These two long two-by-fours will be the top and bottom plates. Place them together, thin edge up, just inside the chalkline (see Fig. 7-16). The wall will be made between the bottom and top plates when they're separated later; when they're tilted up, the bottom plate (B) will sit just where you want it, at the edge of the platform.

Now lay out the plates to show where the studs go; do all the marking at one time to minimize the chances for error. (Since every lumber company sells precut studs of any given length, I'm assuming you're either buying these or making your own from used material. You can buy studs from $88^5/_8$ to $92^5/_8$ inches long, depending on the ceiling height you want.)

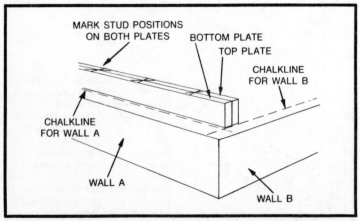

Fig. 7-16. Mark both the top plate and bottom plate at the same time, to insure that the studs are nailed exactly perpendicular between them.

Mark the bottom and top plates together, so the studs can be lined up perpendicular with both plates when nailed. First do the windows, doors, partitions, and corners. Looking at Fig. 7-11, the floor plan, observing wall A from left to right, the first wall part seen is a corner. Since a corner is essentially three studs nailed together in a sandwich, make three marks $1\frac{1}{2}$ inches apart, starting from the end of the plate, since two-by-fours are $1\frac{1}{2}$ inches wide. Mark each of these three spaces with a large × .

Next along wall A, going from left to right, is a window. An actual house plan would indicate how far the center of a window is from the end of the wall. Measure this distance to the window's center, and make a mark. Measure each way from the mark one-half the window width, and make another mark. Add another mark $1\frac{1}{2}$ inches further yet from the center, filling this $1\frac{1}{2}$-inch space with a T, for *trimmer*. Measure $1\frac{1}{2}$ inches beyond this on each side of the window, and make an × , for *trimmer stud*. If the window is larger than 6 feet wide, double trimmers will be necessary. Add an extra trimmer between the first one and the stud next to it. That's all for window layout. In framing, wall openings for doors and windows are *rough* openings. For doors, the rough opening should be the door size plus 2 inches; window openings vary according to the manufacturer.

Partitions are like windows, except they're only $3\frac{9}{16}$ inches wide; measure to their centers, mark $1\frac{7}{32}$ each way and draw a large letter P between the marks.

The next corner is done just like the first, with three × s ($1\frac{1}{2}$ inches wide) marked against the opposite end of the plates.

The section of wall where the plumbing pipes will be must be made with 2- by 6-inch studs, remember; substitute the larger boards if appropriate.

Next the regular studs are laid out. Hook the tape measure over the end of the bottom plate, and pull it out to 16 inches. Make a heavy black mark on the plate at $15\frac{1}{4}$ and $16\frac{3}{4}$ inches, and mark an × in the space. The stud will now be *centered* on 16 inches.

Pull the tape further, until you reach 32 inches—a good tape has special markings every 16 inches—and repeat the process, marking at ¾ inch either way and inserting an × in between. Continue extending the tape, following this pattern to the other corner. There must be a stud marked every 16 inches, measured from the left end, unless something takes its place—a trimmer, partition, or corner, for example. When studs fall inside the space reserved for a window or door, use a C instead of an × ; these will be *cripples*.

Next, take a framing square and transfer all the marks from the bottom to the top plate, making sure that the ends of the bottom and top plates are flush when you begin.

With everything marked, pull the top plate away from the bottom plate, toward the middle of the platform, leaving enough space to fit the studs between the two—about 8 feet. Toenail the bottom plate in position with nails every 6 feet or so. When the wall is tilted up, having the bottom plate secured to the platform in this way will help keep it from slipping. Now insert in the following order: the headers, trimmers, sills, cripples, partitions, corners, and studs. Nail through the plates into the ends of the studs. Many lumber stores will precut trimmers, headers, sills, and cripples for an extra fee; it will save time as long as you can order exactly what you need.

To fit two boards together exactly, first pound the nail through one board so the point of the nail barely emerges from the other side. Then put the two boards together the way you

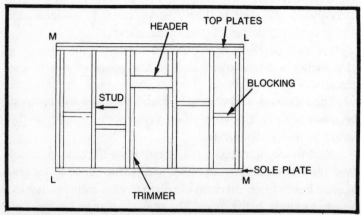

Fig. 7-17. You can be sure that any framework is perfectly even by measuring the diagonals. When a measurement between points L−L equals a measurement between points M−M, the framework is even all around.

want them and set the nail into the second board with one strong blow; driving it in will fasten the boards as you wish.

With the framework nailed together, stretch the measuring tape across each diagonal, from corner to corner (Fig. 7-17); a line drawn from points L to L must be the same length as line drawn from points M to M. Use a sledgehammer to adjust the wall until they are equal. The wall is perfectly square when these lines are exactly the same length. The sheathing can then be applied.

If you have plenty of whole sheets of plywood, the fastest way to put it on is to simply nail the whole sheets on, then cut out the doors and window openings afterwards, before the next wall is started. With the sheathing on, wall C can begin.

Wall C is built just like wall A. When both are completed, it's time to tilt them up and make room for constructing the other two walls, B and D.

A 40-foot wall section is a terribly heavy object. Two men might lift a 10-foot section by themselves with brute strength, but never a 40-foot one. Wall jacks make it possible to tilt the whole thing up at once, however. They rent for a fee of perhaps $1.50 a day. Pry the wall up with a crowbar high enough to slip a chunk of two-by-four underneath one edge. The wall jacks fit under the top plate, just like a bumper jack on a car. It's best to apply the second top plate before you start jacking, since it will strengthen the wall and help keep it from breaking in two if the two jacks aren't equally high.

You should have a jack every 15 feet. When jacking the wall, never stand in the area where it will fall if the jack slips; if you absolutely must, stand in a window area, so if the wall falls back it won't crush you. A 40-foot wall with plywood probably weighs several thousand pounds, and wall jacks are none too reliable.

The jacks work on two-by-fours. Be sure to use two-by-fours with no knots. If the jack inches its way up and reaches a knot, a point of extreme weakness, the board may break, sending the whole thing crashing down. Stay out of the way in any case.

Nail 2- by 4-inch braces into the window opening frames, attaching their other ends to blocks nailed onto the floor. These braces will be all that is holding up the wall until all the walls are connected at their corners and the roof is installed.

After walls A and C are tilted and braced securely, walls B and D are made. The process and layout is the same as for A and C except for one change. Walls B and D are made to fit between the other two walls; they have no corner posts, only a single stud at each end of the wall. The plates are only as long as the space *between* the other walls. They don't hang over the end of the platform, or even reach the end, since the corners of walls A and C do that. Walls B and D are built lying down; when raised into position, their end studs are nailed to the corner posts of the other walls, forming a strong, tight connection.

FRAMING TRICKS

It's a good idea to try to get all the walls as vertical as possible. Make the corner posts vertical with a level, and brace them in position. For each wall the process is the same: Nail a small piece of plywood to the inside of each post top and stretch a string from plywood piece to plywood piece. Then move the wall top all along its length so it's exactly one plywood thickness from the string at all points. Usually, removing a wall brace will make it easy to tilt the wall in or out as needed. But if you have trouble getting enough leverage, nail one end of a two-by-four to the top of the top plate, wide side facing down; nail the other end of the two-by-four to the floor. Now you can move the wall inward by simply bending this new two-by-four down.

The interior walls are simply more combinations of the outside walls. No windows and fewer doors are present, although there are more corners and partitions. There isn't any plywood sheathing inside, so all the walls are quite light, easy for a single person to tilt into place.

If the house is on a stilt-type foundation, the cross bridging or solid bridging (Fig. 7-18) should be installed as soon as any walls are up, if not sooner. I met a contractor who waited until the roof was done before installing cross-bracing. The house toppled over one windy day, because the structure underneath was too wobbly.

118

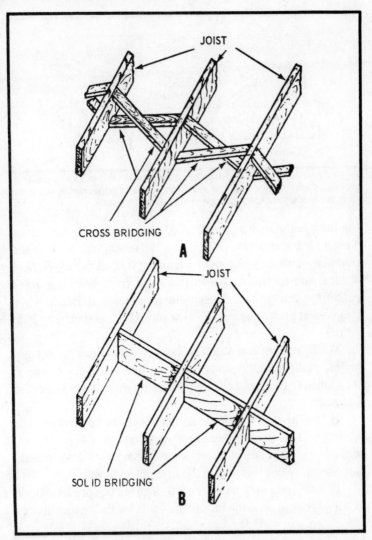

Fig. 7-18. Cross bridging or solid bridging adds extra support between wall studs or floor joists.

All walls are nailed down permanently to the floor by driving one nail through the bottom plate into the floor between each stud pair. Do not remove the bottom plate from door openings until the roof is finished; the bottom plate gives the unfinished walls extra strength in the meantime. If any wall refuses to move into its position on the chalkline drawn on

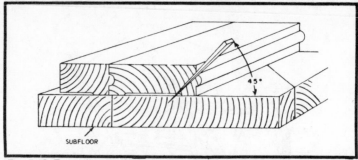

Fig. 7-19. Toenailing, nails driven at an angle, will help keep tongue-and-groove boards together, as well as offering a neater appearance, since the nails are hidden under the groove of the adjoining board.

the floor before it was constructed, you can beat it into place with a sledgehammer. Or you can try toenailing through the sole plate into the floor and banging on it until the movement occurs; the toenailing keeps the wood from bouncing back each time you strike it. Toenailing also comes in handy when tongue-and-groove boards don't want to fit together (see Fig. 7-19).

While nailing the studs in between the top and bottom plates. make sure the studs are all in line with each other; if they aren't. the finished wall covering will bulge over the irregularities.

If the floor under the walls isn't level to begin with, the walls won't be level either. Low spots can be lifted, though, with shims (thin bits of wood) as necessary; the floor should have been leveled this way in the first place, however.

Before the ceiling joists or rafters go on, be sure to include *nailing strips* around the tops of the walls for the ceiling boards

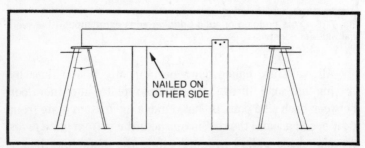

Fig. 7-20. Lifting heavy beams into place is easier if two-by-fours are nailed (temporarily) to the beam and used as guides by helpers.

120

to be attached to; the ceiling boards will be nailed to the underside of the ceiling joists; but at the edges of each sheet, where the ceiling meets the top of the wall, there must be something to attach the edges to.

One problem working above the first floor has to do with hoisting heavy beams into place. One solution is to increase the number of helpers possible by nailing two-by-fours as handles onto the beam between the two ladder climbers, as shown in Fig. 7-20. There also framing accessories to make your job easier (see Figs. 7-21 and 7-22).

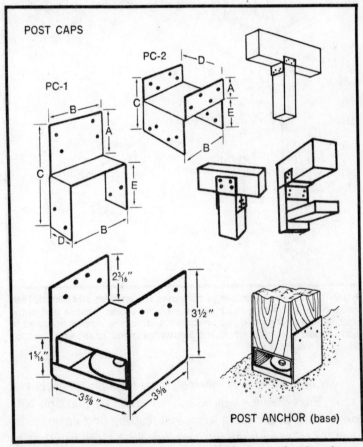

Fig. 7-21. There are special accessories available to make framing a house much easier, even easy enough to be done by one person. Post caps are used to join beams and posts as shown. Post anchors are bolted down and accept the post. (Courtesy Teco Inc., Wash., D. C.)

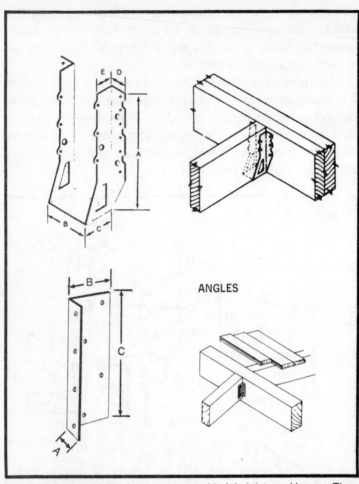

ANGLES

Fig. 7-22. Joist and beam hangers are used to join joists and beams. They make one-man framing much easier, allowing you to rest one end of the joist in the hanger while you position the other end. Angles are used to add support to the spot where two boards are joined, as shown. (Courtesy Teco Inc., Wash., D. C.)

An ordinary power saw only cuts through 2½ inches of wood. Heavy beams can be cut, however, by marking both sides of the beam in the same spot, cutting first on one side, then the other with the saw; a 5-inch-thick piece can then be cut with ease.

Cutting angles, as for slanted walls and roof beams, requires precision, a ticklish task for an ordinary power saw, using marks made with a framing square. But a table saw can save time and anxiety.

Be sure to paint every board that will be exposed *before* installing it, especially boards for the ceiling, deck, or exposed beams.

Whenever decking is used, the individual board ends must be staggered for strength; the amount you need to cut ends, however, can be minimized by buying different lengths directly from the lumber store, or by buying end-matched boards, where the ends are tongue-and-groove.

Roofing

Just after I finished nailing down the floorboards on the second floor, I stumbled upon a newspaper advertisement offering hand-split cedar shakes for half the market price. My inquiry brought me to a place 50 miles away.

I didn't know the first thing about roof shingles or shakes—handmade or otherwise—but the price was right. And I liked the idea of putting handmade shakes on my handmade house.

The place was near a familiar landmark along the freeway and I found it easily. The house looked dreary and weatherbeaten; it was built completely out of cedar shakes. The place looked deserted. I knocked at the door; no answer. I was about to leave when a woman appeared and directed me "out back."

"Jim!" she yelled in that direction. A tall, gaunt, elderly man appeared; I walked over to him at the workshop in back. As I approached, he carefully put down his tool, something resembling a chopped-off meat cleaver with a sideways handle (a *froe*).

"Help you?" he asked softly. He exuded calm, as no business-world person does—a mystical calm. I liked him.

"I need some shakes," I said. He looked straight into my eyes.

"Don't have any now." His face was wrinkled with good times and bad. and a long time in the country weather. "Won't have any until the first of the month." he said. consulting a nearby clipboard.

"That's fine." I answered cheerfully.

"Whew! Sure is cold." he said. smiling. It was the first cold weather of the year. He moved toward his oil-drum furnace. warming his hands. "How many do you need?" His voice was warm; he wasn't a clerk taking an order. but a friendly citizen doing me a service—like a village blacksmith agreeing to shoe a friend's horse.

"Thirteen squares." I answered. A *square* is enough material to cover 100 square feet.

"Okay. mark you down." he said. writing on the clipboard. "for thirteen squares." His hand carefully scrawled out the delivery date and number. Long bony fingers. mottled with enormous knuckles and stringy muscle tendons. made so from years of banging them in the wood-splitting process. did the writing. "Name?" I was looking at the piles of cedar chunks from which he made the shakes. absorbed in imaginings of the years of effort. I didn't hear him speak. "Can I get your name?" he repeated. I finally answered. and he added my name to the other information on the paper.

"You don't deliver. by any chance. do you?" I didn't want to force my tired truck on such a long trip. He said he didn't. maybe I could rent a truck. I thanked him. starting back to the car. He had boards and fencing materials. too. which he pointed out as we walked. all for less than half the lumberyard price; it seemed that he was sacrificing himself. that he could get more for his work than he did. But remembering the days when my work went well. I could understand him liking the easy routine without complicated business matters to interrupt. "Have you been doing this for quite a while?" I asked.

"Twenty-six years." he said. We were on a slight hillside; he pointed to a tract of houses below. "Did all those." he said. just as I noticed all the cedar-shaked roofs. "People come from all over. Had one outfit from way down on the coast come here for shakes on their motel—over a hundred squares. They came and got it in one big truck and trailer."

"That must have taken a long time for you to make," I said.

"Oh, not too long. I get about three squares a day."

Retrieving the shakes a month later came just before the big snowstorm that accompanied my installing the roof deck. I had planned to use my old truck, but it had its own plans. I let a friend use it to move furniture; later that day she called to say it was waiting for me on the freeway off-ramp, refusing to move and still loaded with her furniture. Thinking it was simply being stubborn, out of gas, or something, we pushed it for several blocks, complete with furniture. With Diane behind, pushing with the car, I tried to make it start; it only made sounds like someone blowing his nose. A gas station attendant said it needed a carburetor and a tuneup. Two weeks and $45 later, it carried me three more blocks before it died.

I borrowed a truck from a friend to get the shakes, arriving just as they were ready. I watched the man work for a few minutes, flinching each time he struck the froe to split off another shake. Three squares would certainly take all day; I didn't envy him this knuckle-bruising work. The cedar smelled great, though. He gave me a pencil with his name and motto printed on it: *If it's from cedar, I can get and split it*.

With the roof now waiting for the shakes, I tried to find someone to install them. I had no desire to work on the roof again. I called every listing in the phone book, every ad in all the local newspapers. But cold, icy, winter weather, it appeared, was vacation season for roofers. I did manage to lure two men out to look at the house, but as soon as they saw the steep roof so far up off the ground, they left. Another offered to come out when the weather improved—but only if I paid him three times the standard price. I was preparing to bend to this near blackmail when I got a call from a man *asking* me if he could have the work.

I wondered if he knew what he was doing, asking for such an unpopular assignment. I went over to his house to sign the contract before he changed his mind.

He started working the next day, bringing several ladders, a helper, and some tools I didn't recognize. The frigid weather, especially on the high exposed roof, brought him down often.

This was his first job on a roof like this. He liked working with wood shakes—a welcome change from the asphalt shingles he had used before—but it was going "a little slower" than he expected. I could understand how it might, since his hands, like mine, were freezing; and wearing gloves wouldn't allow him to hold the tiny nails.

I had hoped he would work steadily until the job was finished. But the weather remained cold, his truck broke down, and he had trouble finding a helper. After several days of his first absence, I called him.

"Won't that guy who was with you the first day help?" I asked.

"*Good* help," he said, "I need *good* help. I can get someone day after tomorrow who'll be good help." It sounded funny hearing someone my own age complain about poor help; it seemed like something a parent would say, although I had thought the same thing myself while working on the house. I wasn't sure he was telling the truth anyway; perhaps he didn't plan to return for a long while, and was just giving me a line. I almost volunteered to help him. I thought it over and decided, however, that I wouldn't; I was paying him to do this for me—especially to assume all the worries. I put it out of my mind.

One day the new plumbing inspector paid a visit; he was the nicest person yet to come from city hall. He had some notes the other plumbing people had made about my house.

"That's really too bad about your french drain," he said in a tight-lipped way. "Those guys downtown don't know anything about this hillside." He shook his head, dismayed. "How did you get all the gravel down there?"

"With a wheelbarrow," I said, as coldly as I could.

"Down that path?" I nodded.

"Didn't you use a chute partway?"

"No," I said, shaking my head.

"Oh, boy. That's *really* too bad." After a moment of silence, he said, "Well, we'll see if we can help you from now on." This final confirmation, telling me again that the ditch had been a waste of time, was upsetting. But he sounded very sincere with his offer to help.

"That'd really be nice," I said.

"Okay. As long as I'm here, why don't I write out the permit and get that out of your way." He took some papers from his briefcase. "The cost depends on the number of plumbing fixtures you're gonna have." He pulled out a pencil and consulted one of the sheets of paper. "How many sinks will you have?"

"One kitchen sink, two bathroom sinks," I said. We went down the list: water closets (toilets), bathtub, water heater, dishwasher, washing machine, garbage disposal. The permit came to $65. I expected some very modest fee, like $5, so the price was an unpleasant surprise. But I got my money's worth, since he returned to help me several times. He said I could pay later if I wished, which I did.

"Remember now, any time you have a question—anything at all—just call me," he said. "I'm out here all the time; I can usually drop by the same day."

The time for plumbing hadn't yet arrived, so I went back to working inside the house. The weather remained below freezing, so all the roof leaks were still solid—little icicles hanging all around. The floor was covered with a slippery layer of frozen sawdust and ice; I could have skated on it. My senses didn't know what to make of it; the walls and roof were a house, yet the ice and snowdrifts piling up inside under the windows was disheartening. The outdoors was indoors. My work of the past months seemed nullified; I still didn't have a house.

Working with thick leather gloves, especially trying to hold nails, was an unbelievable hindrance. Other construction chores lay dormant; no contractor could afford to pay standard wages to carpenters for work at this gloved pace. The time was right to do some necessary shopping.

ROOFS

Looking at it from the outside, a slanted roof, like every other part of the house, has several layers: the *outer* protective/decorative layer, and the *inner* skeleton (see Fig. 8-1). The roof is basically a tilted outside wall. The part you see

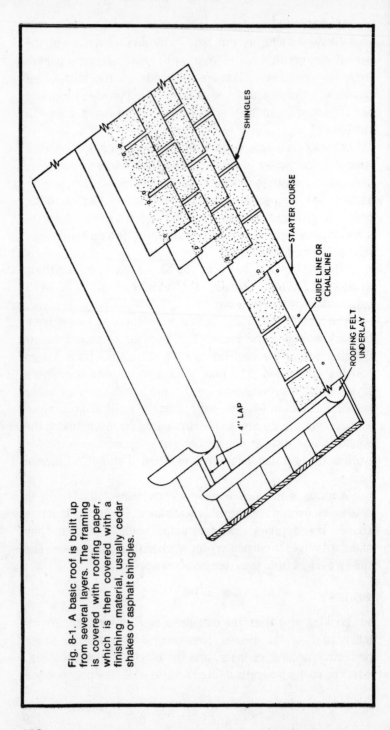

Fig. 8-1. A basic roof is built up from several layers. The framing is covered with roofing paper, which is then covered with a finishing material, usually cedar shakes or asphalt shingles.

SHINGLES

STARTER COURSE

GUIDE LINE OR CHALKLINE

ROOFING FELT UNDERLAY

4" LAP

from outside the house is the shingle layer, which acts like fish scales to let the water run off. Wall siding could be used instead of shingles, except it might leak. Shingles are designed to hold out lots of water, even falling directly on them; some people even use shingles for wall siding.

Shingles

Shingles come in two basic forms: natural wood such as cedar, and synthetic compounds, such as asphalt. Most houses have the asphalt type; they're easier to install, and cost 50—75% less to buy in the first place. Wood is expensive, both to buy and apply; but it looks nice, especially in a wooded setting where the house is meant to blend with the surroundings (see Fig. 8-2). Cedar is a good insulator; it's light in weight and strong, and lasts far longer than asphalt if applied correctly. Some pioneer cabins still have their original cedar shakes. Cedar shakes are roughly split shingles that are

Fig. 8-2. Cedar shakes offer a more attractive finish than does a shingle roof. This view shows our cedar-shingle roof, especially attractive in the wooded setting. As you can see, we also used cedar as a finished siding material.

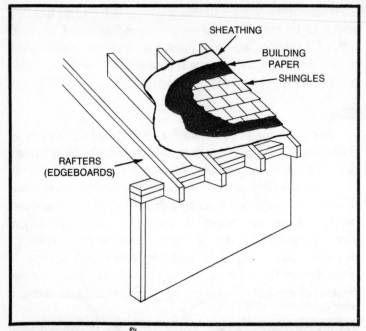

Fig. 8-3. Roofs must be strong enough to hold the weight of people who may walk on them, as the weight of piling snow in the winter time.

much more rustic looking than the smoothly sawn shingles themselves. All are easy to apply without help unless the roof is dangerously steep or otherwise hazardous.

Underneath the top layer is building paper, a second layer of defense against the rain. Building codes require extra-heavy (30-grade) paper for the roof—twice as thick as that used on the walls. Since shingles are expensive, building paper can be used by itself (unless the neighbors object to its appearance) to keep the rain out, as long as it's properly held down with strips of lumber to keep it from tearing in the wind.

Unlike the walls, the roof must be strong enough to withstand wind forces blowing against it and the weight of people walking on it or of piling snow. It must be nearly as strong as the floor, as well as being weatherproof. Once again, as with the floor, walls, and ceilings, edge boards are the skeleton. This time they're called *rafters* (see Fig. 8-3).

You may see some other things besides shingles on the roof. Little pipes sticking up are probably part of the plumbing system. The place where the electricity comes into the house, unless it comes in underground, is often on the roof, in the form of another pipe sticking out of the roof with some wires attached to it. If the kitchen has an exhaust fan, the pipe to vent the air may be on the roof. Heaters and refrigerators powered by natural gas have pipes that come out of the roof. There may also be vents to let air into the attic without letting rain in with it.

Attic Space

On the inside of the rafters is the *attic space*. Some houses have finished attics, where one of the usual finishing materials is nailed across the rafters on the inside as if they were wall studs or ceiling joists. In most houses, however, the inside of the roof is left unfinished, since the slope of the roof isn't great enough to allow anyone to stand up in the attic. In this case, the view from the inside of the attic is of the rafters and the roof sheathing that sits on top of them. Imagine the underside of Fig. 8-3; it looks like the wall sections before the wallboard is put on.

From the living room of the average single-story house, the ceiling you see is not the inside of the roof, but the underside of the attic floor; the inside of the roof is only visible from the attic. Figure 8-4 shows this.

Figure 8-5 focuses on the individual unit of strength for the roof, the *rafter*. Each pair of rafters sit atop the walls, attached to each end of the nearest ceiling joist. The ceiling joists form a link between the lower ends of each rafter pair, and keeps them from collapsing outward. The ceiling joists in this triangular arrangement perform two jobs: they hold the rafters up and they provide a surface for the finished ceiling material.

DESIGNS

Rafter/joist assemblies are available in prebuilt units for roofs of moderate slope, saving time and money. However, these units differ slightly from the simple triangle of Fig. 8-5. They're called *trusses* and look like Fig. 8-6. The added bracing makes the whole thing stronger than a simple

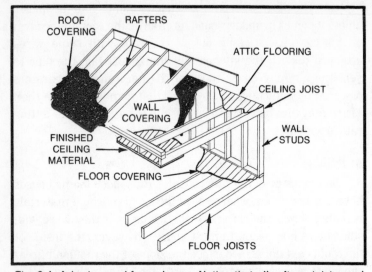

Fig. 8-4. A basic wood-frame house. Notice that all rafters, joists, and studs are placed on 16-inch centers. In this average single-story house, the first-floor ceiling is the underside of the attic floor. The inside of the roof can only be seen from the attic, as opposed to the exposed-beam ceiling that allows the inside of the roof to be seen from the first-floor living space.

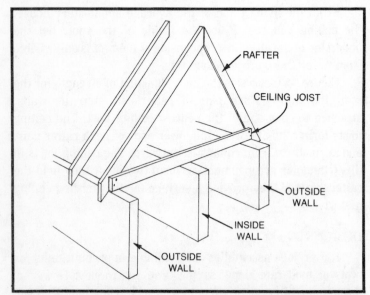

Fig. 8-5. The individual unit of strength for the roof is the rafter. Each end of the ceiling joist is attached (nailed) to the rafter, and the rafter/joist assembly is nailed to the outside wall. Also notice that the inside wall supports the middle section of the joist.

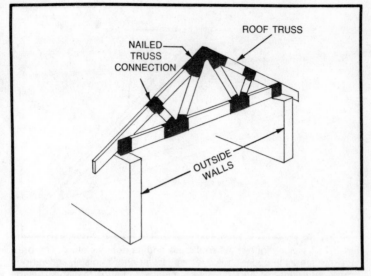

Fig. 8-6. A typical roof truss on a shallow sloped roof. The added braces (truss connections) make this assembly stronger than the simple rafter/joist assembly. The truss bridges from outside wall to outside wall without sagging; no inside walls are necessary to support the middle section of the joist.

rafter/joist triangle, so no inside walls are needed to support the ceiling joists from sagging, as they would otherwise. In houses made with normal rafter/joists, the placement of the inside walls is somewhat preordained since, as mentioned before, the walls must act as midspan supports. Roof trusses, on the other hand, support themselves completely, so the inside walls can go the way most esthetically pleasing. Figure 8-7 shows a typical roof truss and its nomenclature.

Different kinds of roof designs come from modifying the basic structure shown in Fig. 8-4. Leaving out the attic and rafters, while putting the roof sheathing, building paper, and shingles directly on top of the ceiling joists, results in a *flat* roof such as that of Fig. 8-8. The time and cost of making the rafters, the troubles with working on a slanted surface, and the cost of heating wasted attic space are all eliminated, with no loss in usable living space; the attic is just an extra storage area anyway. A flat roof can't use shingles, however; the shingles must be replaced with a special roof coating and sealed with hot tar, since the water won't run off a flat roof as

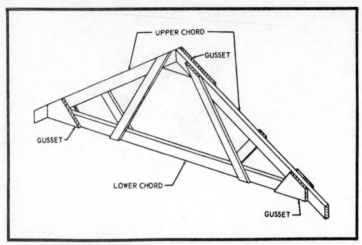

Fig. 8-7. A typical lightweight roof truss and its nomenclature. The principal parts are the upper chord (rafters), lower chord (joists), and various braces and supports known as web members. Notice that this truss is joined at the corner with plywood gussets, but they might also be metal.

well as it will off a sloped roof. In spite of the economy, most houses come with slanted roofs, however; perhaps the box-like look of flat roofs is alien to many people.

Beginning with a flat roof, increasing the height of one of the outside walls and giving the roof a slant creates a *shed* roof (Fig. 8-9), used extensively in contemporary designs. Flat roofs and shed roofs are particularly suited to the inexperienced builder, since they leave out a major time-consuming and exacting step of house construction—

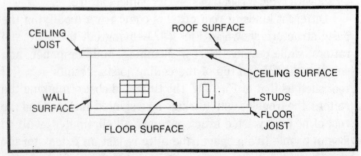

Fig. 8-8. A house with a flat roof has no rafters. The roofing materials are applied directly to the ceiling joists. In the case of a flat roof, however, the final layer must be sealed with hot tar, since water will not run off a flat roof.

136

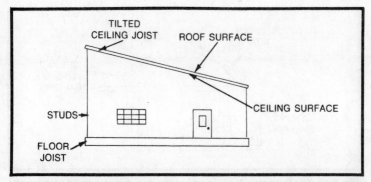

TILTED
CEILING JOIST   ROOF SURFACE

CEILING SURFACE

STUDS→

FLOOR
JOIST

Fig. 8-9. Increasing the height of one side of a flat roof produces a shed roof, used extensively in contemporary designs. This design is particularly suited to the inexperienced builder, since it eliminates the time-consuming and exacting step of installing rafters, and decorative shingles or shakes can be used.

cutting and installing rafters. If you want to save time but prefer a normal slanted roof with a peak, you can use a design relying on a *ridge* beam.

The illustrations that follow show the basic steps of building a ridge-beam roof. In Fig. 8-10, we start with a simple

Fig. 8-10. If the traditional sloped roof is desired, much time and labor can be saved by building a ridge-beam roof. The first step is to begin with a simple house—no ceiling or roof.

house, no ceiling or roof. Figure 8-11 shows the ridge beam, a long and heavy beam across the empty top of the building. This beam is supported at each end by posts projecting up and out

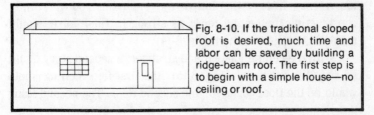

Fig. 8-11. The second step in installing a ridge-beam roof is to add the ridge beam, a long, strong, and heavy beam bridging the building. The beam is supported on both ends by posts projecting up and out of the wall. Notice that there is still no roof or ceiling.

RIDGE BEAM

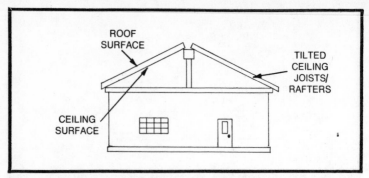

Fig. 8-12. The third step in building a ridge-beam roof is to tilt up the ceiling joists to rest against the ridge beam. The two sides of the roof are finished as though two shed-roof sections were added.

of the wall. A double-shed roof is installed (Fig. 8-12) by tilting up the ceiling joists, as if this ridge beam itself were a wall. In Fig. 8-13, the walls are built up on both sides so they reach the peak of the roof. Presto! The ceiling edge boards double as rafters, saving time and materials; and the tilted roof leaves an open and ventilated ceiling below.

The steeply sloped roof in Fig. 8-4 leaves enough headroom for a cozy little attic, hideaway, or living room; simply constructing the ceiling joists as strong as floor joists would make this transformation (Fig. 8-14).

With the ceiling joists made stronger, a second story could be added to the house (Fig. 8-15); the first-floor ceiling joists would be the floor joists for the second floor. The second-floor

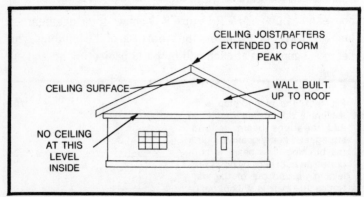

Fig. 8-13. The final step in building a ridge-beam roof is to build up both side walls to the top of the peak.

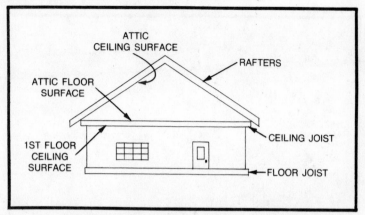

Fig. 8-14. A steeply sloped roof leaves enough headroom for an attic, hideaway, den, etc. Simply use ceiling joists that are as strong as floor joists, in order to support the first-floor ceiling and attic floor.

walls, ceiling, and roof would then be installed in the same manner as for the first floor.

Buildings with more than one level are, in fact, made this way, rising as if by bootstraps. Each time a floor is built, the walls go up as if this were the first floor; when the ceiling joists are placed, they become the floor joists for the next level floor, and so on until the rafters follow the ceiling joists instead of another finished floor.

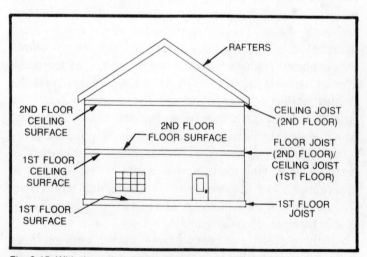

Fig. 8-15. With the ceiling joists made stronger to support an upper living space, a second story could be added.

139

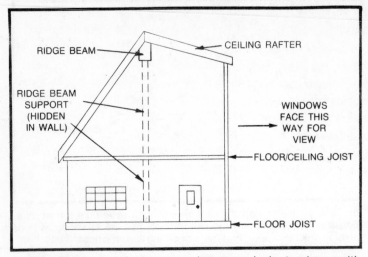

RIDGE BEAM

CEILING RAFTER

RIDGE BEAM SUPPORT (HIDDEN IN WALL)

WINDOWS FACE THIS WAY FOR VIEW

FLOOR/CEILING JOIST

FLOOR JOIST

Fig. 8-16. The house we built is a cross between a single-story house with an attic and a two-story house—a 1½ story house, using a ridge beam.

My house is a cross between a living-area attic and a two-story house, using a ridge beam (see Fig. 8-16).

EXPOSED BEAMS

You can have exposed beams (see Fig. 8-17) if you keep the attic area but omit the attic floor and finished ceiling material. The ceiling joists and rafters become visible from within the house. Leaving the framing exposed may seem unappealing at first, but it's essentially the way to obtain exposed-beam ceilings. Exposed-beam ceilings, as the name implies, however, aren't the same as simply leaving off the finished ceiling to expose the underlying edge boards. The edge boards we've discussed so far are two-by-sixes and two-by-eights; by no means could you call them *beams*. Even if you could, they're spaced so close together, 16—24 inches on center, that leaving them in view would create a cluttered appearance. For exposed-beam ceilings, the underlying framing must be made heavier. And more widely spaced edge boards (*beams*) are required.

If there aren't any heavy beams around, perhaps three joists or rafters nailed together to form a single large beam would work. It might not look as rustic as the real thing, but the cluttered look would certainly be minimized. Imagine doing just that in Fig. 8-4. Rafters would be consolidated into two large beams, but the roof material would than sag between them, requiring that it be made of thicker single-board planking. If this thicker roof covering were made from nice looking boards, however, letting them show through would be a pleasure. Presto again! The clutter is gone and the unsightly underside of the plywood roofing is replaced by finished quality boards. (The ceiling joists can't be removed unless the weight of the rafters is taken by a ridge beam.)

You've just discovered *post-and-beam* construction. You can buy the heavy beams, so there is really no need to nail smaller joists together.

Post-and-beam construction can replace not only smaller rafters, but ceiling joists, floor joists, and studs as well; several edge boards can be nailed together, and the resulting

Fig. 8-17. Exposed beams are an attractive alternative to an extra living or storage space. This is accomplished by omitting the flooring and ceiling materials that would normally cover the joists. The inside of the roof, however, should be covered with some kind of finishing material for a more attractive appearance.

larger edge boards can be spaced further apart. Just be sure that the sheathing covering this wide-gap framing is thickened, so it won't sag or bend between the edge boards.

In post-and-beam construction every ceiling joist rests directly on a stud (*post*) below it, as does every rafter. In the regular closely splaced stud/joist/rafter method, the ceiling joist or rafter can sit anywhere along the top plate of the wall; the wall is thus a weight-bearing wall along its entire length.

In post-and-beam construction the studs carry all the weight of the roof and the upper floors, while the spaces between them carry no weight. *Bridges* (*headers*) aren't required where windows, doors, and other openings can fit within these spaces, and floor-to-ceiling windows can be placed in almost every space. The main use for post-and-beam construction is in giving such variations to house design.

Advantages

The main practical advantage, other than appearance, of exposed-beam ceilings is that all the ceiling material, the *roof decking*, is installed from above, prepainted, eliminating the need for neck-craning ceiling installation. The savings from not needing ceiling board may seem worthwhile, especially since it lets you do the work yourself; but the exposed-beam framing is more expensive than the conventional ceiling, since the finished quality and increased thickness of the single-board roof decking (*sheathing* is called *decking* when it's made of single boards) is much more expensive than the thin plywood you could use with narrower spacing. Also, the single boards take far longer to install than do 4- by 8-foot plywood sheets. And, finally, some people use the ceiling wallboard anyway between the beams. Hiding electric wires and plumbing pipes in exposed-frame ceilings is difficult.

Using post-and-beam framing on flat, shed, or ridge-beam roofs will provide exposed beams. The lower level of a two-story house can have exposed-beam ceilings if the ceiling joists are made according to this method.

Installation

Actually, assembling the roof and ceiling section, even for a high open-beam ceiling, is something one person can do. The joists and rafter placement is the hardest part. Two men can lift the beams quite easily.

The first-floor ceiling joists in my house are 20-foot four-by-eights. I installed a few by myself, working one end then the other; it was time-consuming. The rest were easily installed with another person helping. The 18-foot four-by-six rafters went into place against the ridge beam with my sole efforts. The few large rafter/joist triangle assemblies (see Fig. 8-18), where no ridge beam existed, were built lying down then tilted up with wall jacks. I found that any time a jack was helpful, though, it took at least two people to do the work.

I installed all the roof decking alone. It was a one-person task, actually, although another person could have helped save time by working the other side of the roof. I pulled each board up, after leaning it against the wall where I could reach it from the roof, and slipped it into place, its *groove* over the *tongue* of the previous board.   An unusual hazard in building my

Fig. 8-18. The rafter/joist triangle assemblies are built lying down, then lifted into place with wall jacks.

Fig. 8-19. A steep, high roof is a potential hazard. Work from inside, if possible, on a secure ladder.

house was the steep, high roof, as in Fig. 8-19. I managed to do all but the roof peak from the inside on a ladder. Yet even so, working at a height of 20—30 feet would have meant disaster if I had slipped.

The building inspector related a gruesome story of a young man nearby who, in the "prime of life," was building his own house, with an equally steep, high roof. He slipped one day, breaking his back, becoming paralyzed forever. No amount of money saved or pleasing interior could compensate for such a calamity.

I worked on the roof, taking all the safety precautions I could think of; yet I still had the nagging worry that something would happen (e.g. a broken rope), and I'd lose my balance. I wished many times that I had a nice shallow roof.

VARIATIONS

The basic house is described; all that remains is the plumbing, heating, and wiring. As you can see, an average house is basically this double-layer skin-and-bone affair. The floor, walls, ceiling, and roof are made up of a "skeleton" of edge boards covered with a "skin" of finishing material. The basic shape can vary as the skeleton determines. Different roof styles modify the appearance of a house immensely. Ceilings, walls, and windows change dramatically from post-and-beam to regular construction.

Depending on the finishing material, the house can look a thousand different ways: *rustic* with wooden shakes, *urban* with asphalt shingles; *formal* with brick or stucco, *relaxed* with wood siding; *warm* with carpeting, *cool* with wooden floors; *cozy* with wood paneling, *bright* with painted wallboards, and so on.

Enclosing the House

9

With the temperature so low, the fireplace was my next order of business. Some months before, I stopped at a half-built house to ask the contractor where he bought his prefabricated fireplace. (I knew he was the contractor because I passed the house several times before and saw him standing around—*supervising*.) He hesitated for a moment, as if I had asked a trade secret, but told me where he bought it.

"They give me forty off," he said, which evidently meant a 40% discount.

"How do you get that?" I asked. He asked me if I were a contractor. "No," I answered, "but I'm building my own house."

"Well, maybe you can give them a story or something. You might as well try it." He gave me the address; I thanked him and left.

The *showroom*, as a wholesale distributor's display area is called, was in the industrial part of town. I went there the next day. A forest of bright metal fireplaces greeted me on opening the door. Several desks-with-secretaries and a few salesmen populated the surrounding area. The room was half store, half office. I wore clean jeans and a flannel jacket—what I thought would be contractor's garb—to look official as I gave my "story." I spotted the unit I wanted and, soon, a secretary came over, a welcome relief from the sweaty, dusty hillside I had just left. I told her I wanted "one of those," pointing to the model I liked.

"Our standard discount," she said, "Is thirty percent off the list price. If you'll be paying cash, you get another two percent for that."

She sounded like she was doing me a favor. I remembered the 40% the contractor had mentioned, however. I looked at the price on the list, something around $400.

"A friend of mine said I could get forty off," I said, purposely omitting the "percent" for professional effect.

"That's for contractors. Are you a contractor?" Couldn't she see that my outfit and slang made me one?

"Well, I'm building my first house right now," I said. She looked at me a moment.

"Well, we have to send a salesman out to the house to check the order exactly anyway; I'll let you work that out with him."

A few days later a salesman appeared on my then roofless deck.

"Boy, you really want to get away from it all, don't you," he exclaimed, reaching the bottom of the path. "Quite a place you have here!"

"Thanks."

"This is your first one, I understand." he said.

"Right."

"You know, I have a friend who builds houses like these—secluded, woodsy. He makes them for himself, but someone always comes along and offers him three times its cost and he sells it. His wife is pretty mad at him right now.

He told me about his house up in the mountains; he liked to ski. but the heating bills were so high he spent every weekend chopping wood for the fire—which brought us to the topic at hand.

"We're not supposed to give this forty-off discount unless you plan to build six houses this year. We make you a distributor then. Do you think you'll get six done this year?"

"Well, I don't know," I said. "This is the extent of my work so far." I slapped the wall. "And I don't know." I smiled, nodding. "I might."

He chuckled and said, "That's right, you never know." We both knew I wasn't going to build any more houses. "I'll just call you B. C. Construction and give you the discount."

Perhaps he thought I *might* build more houses some day and, if I did, his help now would be remembered later. I signed the order, thanked him, and agreed to pick the fireplace up myself.

I rushed to install the fireplace, nose red-cold, mustache icicled, hands freezing, hoping to melt some of the snow inside the house. My friend Jack and I carried the bulky thing down the hill, imagining the whole house warming with the glow from the fire.

The roofer's hammering kept me company as I worked; at last the fireplace was assembled. Three days of drying out snow-covered wood produced a disappointing fire, though, that only kept us warm if we stood directly in front of it.

There we stood, captives of this everlasting building process, frozen everywhere except our "fire" sides.

FIREPLACES

The fireplace sits on its own concrete foundation. When the walls are built, the place where the fireplace is to be located looks identical to the other openings in the walls (Fig. 9-1).

If you examine the place where the fireplace is built, you'll see a box made of smoke-blackened bricks. Actually, the bricks are several layers thick and were once clean, just like the ones you see in buildings. A fireplace is nothing more than a small box where the fire is made, with a chimney to let out the smoke. There is a delicate trick, however, to making fireplaces and chimneys properly, so that the smoke goes out by itself; some homemade fireplaces, I've heard, let some of the smoke out into the room due to poor design and construction.

Fireplaces come in many different shapes and sizes, all with the basic box and chimney. They fall roughly into three categories, each with an enormous range of variations: regular brick, freestanding metal, and imitation brick (metal).

Fig. 9-1. The metal fireplace sits inside the framed-in wall. Insulation is placed between the studs, covered by a brick facade.

Brick is often added around a fireplace to make it more massive and hearthlike; a house with a fireplace that seems to cover a whole wall is the extreme of this idea. Masonry is hard, exacting work, and freestanding metal fireplaces eliminate the need for it. For those who don't appreciate the metal creations, however, the imitation models allow you to have a metal fireplace and hide it too (see Fig. 9-2).

WINDOWS AND DOORS

Back when I was excited about finding used lumber, I noticed an advertisement for used windows. The company was removing a whole neighborhood of houses from a freeway construction path about 30 miles away. The first time I went there, no one met me at the designated time and place—a 60-mile drive for nothing.

Fig. 9-2. The fireplace in the finished house. The metal firebox has been decorated with a brick facade and hearth.

"You'll have to come out and see what we have," insisted the telephone voice. "I can't tell you over the phone."

I returned; again no one could be found. The outfit seemed to be a fly-by-night organization. I found a house with small window panes, as I wanted, but they didn't open with a crank; screens would be hard to install. I made a few more trips, though, finally contacting the person in charge. But I couldn't be sure these used windows would work; Big Dee's comment about banks not liking used material came to mind. So, after all the effort of finding something, I gave up.

I didn't order new windows in advance, as I had the fireplace. I thought windows were a basic item that I could walk into a store and carry home the same day. The little study I'd given to window pricing hadn't led me to believe the contrary. Wooden windows cost two to three times the price of aluminum windows. Insulated glass costs twice as much as normal glass.

But the clerk said a window order would take two or three weeks to fill. The problem was that my window sizes were all unique and had to be custom-made. Designing windows as I went along had been foolish; I should have found out what sizes are common and used them. I designed my windows to be 3.5 by 4 feet. Stock size would have been 3 by 4 feet—no great difference in size, but saving time by allowing me to buy off-the-shelf items.

Buying aluminum windows was an unpleasant experience. The delay had me in bad enough spirits, but the salesmen irritated me completely. From pricing inquiries, I had settled on a certain large factory; but something about the sales office gave me a chill, and I would find out later why. The three salesmen were young, aggressive, hard-dealing businessmen. For a simple question I had about price comparisons, the clerk refused to let me see his price sheet. I looked over the counter, reading upside down. He pulled it back, covered it with another paper, and glared. I wanted to get the windows ordered, so I had him add up the total sum for my windows and signed the order; he said he'd call when everything was ready.

I had left one window out of this order—the large living room window facing the street. It would be wood-framed, with divided panes to enhance the cottage effect. I visited a small company nearby, inquiring about this window. Plaques covered the walls, announcing that "Benny's Bashers" had won this Little League pennant or that bowling trophy.

"Sure, we can fix you up," said Benny. He looked through his lists and came up with a price for the window.

"Does that come complete with everything?" I asked.

"Sure does," he said. "Sash, glazing, hardware, and casing." I didn't know any of the terms he used, though, and repeated the question in my own words.

"Is that the frame and everything?"

"Sash, hardware, casing...just slip it in the opening and it's ready to go." His relaxed manner was a surprise after my previous encounter, a welcomed contrast. I ordered the window and a front door as well.

Armed with a new store of plumbing facts, I went home to try and make sense out of them. Before I got very far, though, the window people called saying the windows were ready and I could come and pay for them any time. Then the roofer called, saying he was ready to be paid.

I strolled into the window factory, happy at finally getting these windows. I started writing out the check for the price quoted originally. He gave me the bill, though, and it was $100 over the original quote. My blood boiled!

"What's the deal!" I said. "How did the price go up?" The smart young man, in snappy clothes, was composed.

"Up? Oh, I tried to call you after you left. I forgot to include three windows on that first price I gave you." He had only given me a total price originally; I had no way of telling if something was left out, since none were listed individually. I was indeed over a barrel now; I couldn't start shopping all over again. If I wanted to finish the house soon, I had to take the windows and pay this bandit. I wrote the check in silent rage and left, receipt in hand. Low prices indeed! What a lousy trick!

My rage subsided soon, however. I learned by now to let things beyond my control go their way. I had found that housebuilding consisted of working within the limits imposed by chipping away at Mother Nature and Lady Luck; I had long since ceased to expect things to obey my needs.

Back at the house, the roof was done; only windows and doors were lacking. I studied plumbing books for a few days until the aluminum windows arrived.

They came in a delivery truck. The path was muddy for the dozen or so trips it took to carry the glass down. The truckdriver helped, though. Evidently accustomed to carrying fragile items, he carried three down at a time—as I held my breath.

Diane and I spent the whole weekend installing them, amidst much swearing and frustration as we tried to get them leveled. One of us steadied the window as the other nailed—it was a two-person process. But Sunday night we could, at last, stand at a window opening and not have the wind and rain lashing at our faces; the house was enclosed.

We celebrated when we got the building permit; now we just sighed with relief at this new milestone. Making first-floor walls all in one week had been a misleading gage of things to come; visible progress came slower and slower. Installing windows did make a dramatic difference, however; the house was now sealed and we could even live there if we wanted to. But the framework walls inside reminded me that there was so much left to do.

The next day brought the large window for the living room. The factory-made aluminum windows had come in a delivery van. But a man from the small window company brought this window in a pickup truck.

"You'll help me carry this down, I hope," I said to the driver, watching him gnaw on a cigar. He surveyed the twisting muddy path to the house below and nodded, as if his thoughts read, "Well, there's probably a fifty-fifty chance of making it." We hefted it up and out of the truck and carried it down the path. It was heavy, several hundred pounds. I had no help coming to assist me in installing the window, so I asked him if he'd help.

"You wouldn't want to help me put this in, would you?" I said, as we got to the house. I thought we could just set it in place in the rough opening.

"Have you got the opening framed right?" he barked.

"Yeh, I think so." He led us to the window opening and set it on the edge—the window was a little fat. We strained and shoved, but it refused to go in past the half-inch it hung on by.

"Well, thanks anway," I said. "I guess I can get it by myself."

"Let's see," he said. "Got a handsaw?" I fetched one. He rapidly sawed at one corner of the wall that appeared to be catching. We shoved, but the window went in only slightly farther. He grabbed a hammer laying nearby; I cringed. He banged on one corner, then the other, as the window went in only a little more. Using the saw, he attacked another spot where I had goofed on the opening. Still the window wouldn't budge. He looked around, biting his ragged cigar, and spied a sledgehammer. "Hold the window," he said. I closed my eyes and waited for the crash. He walloped the window again and again. It moved into position!

"Hey, that looks pretty good!" I said. He examined the fit.

"Here," he pointed to some tarpaper wadded up between the wall and the window, "You got some paper that needs to come off," he said. He took out a pocket knife, sliced the guilty portion and pulled it off. "Now it ought to work." He picked up the sledgehammer and struck the window frame a few more blows, then gently tapped it as he walked around examining the window. It was perfect. "That looks pretty good." His gruff voice had a satisfied tone. He looked around, taking in the house. "Did all this yourself, eh?" he said, an admiring crack appearing in his tightly closed lips.

"Yep." I said.

"Well, good luck to ya."

"Thanks a lot." I said.

The windows, doors, and other wall parts are built elsewhere, and only later do they become part of the house. Good carpenters can make them on the building site; those with less experience, like myself, can buy them from a door and window company *prebuilt*—thereby gaining expert

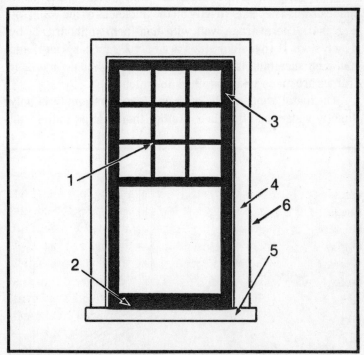

Fig. 9-3. The basic part of a wood casement window: mullions (1), horizontal part of sash (2), vertical part of sash (3), vertical part of frame (4), horizontal frame and sill (5), and casing (6). Incidentally, window parts should be painted in the order listed here.

craftsmanship when you have none, as well as saving lots of time.

Looking at the window, the first thing to note is the glass. *Glazing* is the construction industry term for window glass. A window is *glazed* if the glass is in it. You can buy windows glazed or unglazed (with or without glass).

The glass may look as if it goes directly into the wall at its edges, particularly if the window isn't meant to open. But the glass sits in another window, the part that holds it in place, called the *sash*. When you open a window, the part that moves is the sash. The part that doesn't move is called the *window frame*. In a window that doesn't open, the sash and frame are the same thing. The frame may look as if it's part of the wall; but it was built as part of the window, then slipped into the opening in the wall for the window (Fig. 9-3).

The thickness of the frame for both windows and doors made of wood depends strictly on the thickness of the wall (see Fig. 9-4). The average wall with half-inch wallboard, 2- by 4-inch studs (two-by-four studs are only 3½ inches wide), and half-inch sheathing is 4½ inches thick; the window and door frames are made 4½ inches thick to match.

On metal windows the sash-and-frame assembly is only slightly wider than the glass. Rather than resting in the wall

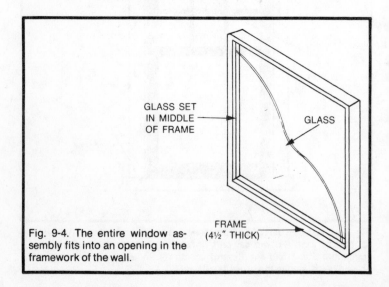

GLASS SET IN MIDDLE OF FRAME

GLASS

FRAME (4½" THICK)

Fig. 9-4. The entire window assembly fits into an opening in the framework of the wall.

opening itself. the metal frames are nailed to the outside edge of the opening.

Some windows have molding around them on the wall surface, like a picture frame, called *casing*. It goes on after the window is put in place to give a finished appearance (Fig. 9-5).

Don't worry about remembering these definitions now. It doesn't matter what anything is called as long as you can put the house together. But it helps to know these words when you go to the store, even if only to speak to sales personnel. They get very annoyed with someone who comes in and asks about the "gadget" or the "whatchamacallit."

Windows have other parts that make up the whole assembly of the framed glass. If you make your own frames, most glass companies will gladly install the glass (*glazing*) for you if you buy the glass from them. Making windows is precision work; you must have a table saw and the proper materials. A friend of mine made his own nonopening

Fig. 9-5. Once the window is set in the wall frame, it is finished with casing to give a pleasing appearance.

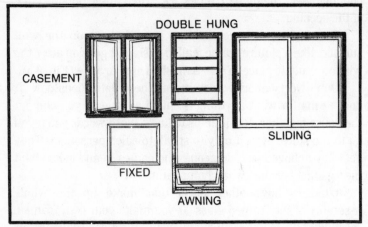

Fig. 9-6. There are many styles and types of windows, their choice depending on your needs and preferences. The windows shown here are the most common of the several types available. (Courtesy Anderson Corp., Bayport, Minn.)

windows, claiming it was fun; but he bought the ones that had to open. Windows come in several different common styles (see Fig. 9-6), available from several window manufacturers. Windows can be custom-made for special applications, but it is better to use standard sizes (see Fig. 9-7).

I was very apprehensive, at first, about carrying glazed windows around, thinking they were very fragile. But a friend who once worked on a window factory's installation crew said their motto was: *If it doesn't fit, get a hammer; if it still doesn't fit, get a bigger hammer.* So I stopped being so apprehensive.

Installation

Installing windows requires two people—unless it's a picture window—one to hold it in place and one to nail it. Wood-framed windows that one person can lift easily can be set in the opening, leveled, and nailed in place without help. Even so, the leveling goes much faster with one person to move the window while the other checks the level.

The fastest way to level a window is to hold the window steady while the other person sticks some wood in the appropriate places to prop it.

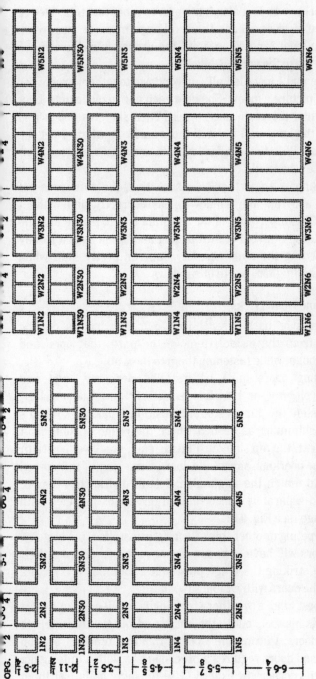

Fig. 9-7. Windows come in standard sizes to fit easily in framed-in walls. The wood casement windows shown here are common sizes found in most supply houses. Note that the dimensions given are for window openings in rough framing. Some manufacturers size their windows and some don't. When framing the rough opening for a window or door, be sure this information is clear. (Courtesy Anderson Corp., Bayport, Minn.)

159

If you have to do it alone, the process goes like this: Wedge in a piece of wood, check the level; use more or less wood, check the level; and so on, until the proper bits of wood are placed to make the window level.

Diane and I installed all our windows together, except for the large picture window. She held each one in place until I got the first nail driven in one corner. Then she steadied it and watched the level as I pivoted the window to where I set the second nail. With the window level, she relaxed and steadied it as I set the other nails.

Leveling is important to insure that the rain runs off properly and that the window opens correctly and without sticking. The importance of the assistant is is to insure that the window doesn't fall out of the opening and break as you jockey around setting the nails. Anxiety can run pretty high when you're installing $500 worth of windows.

Window and Door Tricks

Upstairs or hard-to-reach-from-outside windows can be installed from the inside. Remove the parts that open and reach through, while fastening them to the wall.

Prehung doors must be installed perfectly level and vertical (*plumb*), or they'll open poorly and look terrible. First, fasten the hinge side in position, making sure it's vertical, shimming as needed where the rough opening isn't plumb. Next, the top side must be fastened in a level position; finally, the doorknob side of the door frame (jamb) is fastened vertical to match the hinge side. Install doorknobs—they're purchased separately—according to directions included with the package (see Fig. 9-8).

If carpeting or other thick flooring is installed, the bottoms of the doors will have to be cut. Remove the doors from their hinges by striking the hinge pins with a hammer and chisel, then cut the doors with a table saw.

The best kind of weather-stripping is applied by an expert who brings his tools out to the house and fits the materials to exterior doors. Putting the wooden molding (casing) around windows and doors requires a precision saw with a miter gage to make the corners fit exactly. Use finishing nails; set them

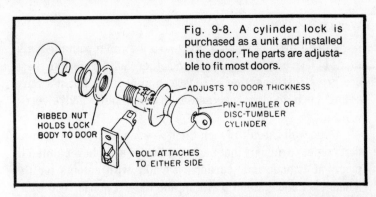

Fig. 9-8. A cylinder lock is purchased as a unit and installed in the door. The parts are adjustable to fit most doors.

ADJUSTS TO DOOR THICKNESS

PIN-TUMBLER OR DISC-TUMBLER CYLINDER

RIBBED NUT HOLDS LOCK BODY TO DOOR

BOLT ATTACHES TO EITHER SIDE

with a nail set, and cover the hole with putty to match the wood. All traces of patchwork will disappear upon application of final stain or paint.

DOORS

Doors are much like windows. There's a moving part (the door) and a fixed frame that the door attaches to.

Doors range in price tremendously, depending on what you want. There are two basic kinds of doors: *hollow core* and *solid core*. Solid-core doors are solid wood. They're normally used for exterior doors, since they're strong and good insulators. They range in price from $50 to thousands of dollars, depending on the design. Prehung doors come preassembled in the frame, with hinges in place, ready to slip into the rough opening in the wall.

Doors used inside the house (interior doors) are the cheaper hollow-core doors, which are really thin closed boxes. You could smash your fist right through one of these if you hit the right spot. They cost as little as $20 prehung, including the doorknobs.

Solid-core doors can cost several hundred dollars if the wood is heavy and the doorknobs are fancy. Inexpensive doors are perfectly good for the interior, although they don't stop room-to-room noise as well as the heavier doors. But they last as long as the others if properly finished, and do what they're supposed to do—create visual barriers between room areas.

Be sure, when buying doors, that the door is not warped; if it is, it won't ever close properly. Door manufacturers will sell directly to you wholesale.

WALLS

When I think of walls, I imagine a log cabin with the walls made of logs piled one on top of the other. Solid walls like these are great for insulation, strength, and beauty; but it's hard hiding such modern necessities as plumbing and electric wiring.

If the walls aren't 10-inch-thick wooden logs, the next thing that comes to mind is that perhaps they're somehow painted in place; if I made four cardboard walls, with cutouts for the windows, I could stand them up and have a house. The walls are made in layers like the floor; the walls you see and touch are the decoration on the inner framing that holds things up.

Materials

The layer you see, the finished layer, can be either *wallboard*, *plaster*, or *wood* paneling, depending on your preference and finances.

Wallboard is a sheet of plaster, coated on one side with a thick layer of paper. It breaks easily along any line where the paper is cut, or it can be cut with a saw. Wallboard is probably the most common finishing material today. It's usually the smooth material under a painted or wallpapered wall. It comes in standard sheets, and it's *very* heavy (even the small sheets are heavy!).

To get a finished wall using wallboard, the sheets are nailed in place and plaster is applied to the seams; the wall is then ready for paint or wallpaper. The material itself is very inexpensive, but it requires an expert touch to cut it and to fill in and finish the cracks; inexperienced workers should start in a closet, where their practice mishaps won't be obvious.

I know one contractor who, after cutting his own wallboard and nailing it in place, swore he'd never touch the material again—just the nailing! The hardest part is sealing the joints between the sheets.

Plaster is much heavier and more expensive, although it's more sound-deadening. Plaster is applied by a highly skilled professional plasterer over a screen-like material called *metal lath* nailed to the wall frame. Getting it smooth—even getting it to stick—is an art. It was used more widely before wallboard was invented.

162

Wood paneling can be purchased in two standard sizes: 4-by 8-foot sheets, like plywood or wallboard, and in single boards. The sheets are the fastest to install, although they require two people to apply—one to hold the sheet in place and one to nail it. Neither kind of paneling requires intricate skill, and though the raw materials themselves are more expensive than the wallboard sheets, they don't need any finishing after installation. Even single boards, the most expensive wall material to buy, is no more costly (if you do the work yourself) than wallboard done for you by experts (which you even have to paint afterwards). Figure 9-9 shows our basement finished with single-board panels.

Installation

The finished layer, whatever it's composed of, is *nailed* or *glued* directly to the wall studs.

Fig. 9-9. We chose single panels as the means of paneling our home, as opposed to standard 4- by 8-foot sheets. Although the wood is more expensive than the standard sheets of paneling, the method is faster and cheaper in the long run. One person can easily install the boards and no finishing or molding is required.

Fig. 9-10. The lower level of the house in the framing stage. These rough-in studs will later be covered with paneling or drywall.

Although walls don't have the weight of people and furniture bearing down on them to make them sag, as do floors, winds and weather push on them sideways to make them bend. Therefore, the wall needs edge boards (*studs*), turned with edges against the wind, to keep the wall from bending. Almost always, two-by-fours are used for studs. These edge boards in walls have one other important function as part of house framing: they hold up the roof (and upper floors, if any). Figure 9-10 shows the lower level of our house in the framing stage.

Siding

Wall studs are covered on the outside, giving them material on both sides. Exterior siding is more than decorative; it provides a layer of armor for the house, serving to let the water run off and protect the house against the weather.

The different kinds of siding you can obtain depend on your location and taste. Lumber siding is quite common in the Pacific Northwest; it comes in plywood sheets or single boards and is easy for unskilled persons to apply. Recently, other materials have been used to simulate single-board siding: aluminum, plastic, and gypsum are becoming less expensive than real wood as time goes on.

Brick and stucco are common in areas far from lumber sources. Stucco is a kind of outdoor plaster, requiring an expert touch to apply. Bricks are another difficult art to master, too, and demand hard physical labor to apply, as well as *brickwork* requiring a special concrete foundation. We used cedar shakes for the siding as well as for the roof (Fig. 9-11).

Between the finished siding and the wall studs are several layers of insulating materials. *Building paper* is paper treated with asphalt to make it waterproof and insectproof. *Tarpaper* is the second layer under the siding, required by the *uniform building code*. It is stapled onto the next layer, the *sheathing*. The sheathing can be either 1-inch boards or plywood. Plywood

Fig. 9-11. We used cedar shakes for the exterior siding, as well as for the roof; but siding can be any one of several materials, including plywood, brick, stucco, aluminum, etc.

is much, much faster to work with, and is stronger and cheaper. The sheathing provides still another layer of weatherproofing and insulation, and increases the strength of the wall.

The sheathing layer may be omitted, with the building paper being fastened directly to the studs. However, it's then necessary to brace the frame diagonally to keep the studs from moving, as the plywood would have done. Little time, if any, can be saved, since the bracing must be done precisely in order to be strong. The house won't be as quiet and as solid inside unless heavy siding, like brick or stucco, is used. With the building paper attached directly to the studs, putting on the siding will be much more difficult; you'll have to find the studs to drive nails into and the paper will hide them from view.

Insulation usually fills the spaces between studs, helping to keep the indoor temperature constant. Plumbing pipes and electrical wires run through this space also.

CEILINGS

Like the floor and walls, an average ceiling is a two-layer sandwich; the part you see and touch is the finished cover on the framing underneath. This finished layer is, like the walls, most often wallboard or plaster. Sometimes special acoustic tiles, which come in 1-foot squares, often with many tiny holes, are used instead of plaster or wallboard for soundproofing. Wood ceilings are another alternative.

The alternatives for finished materials used for walls and ceilings are the same; but the installation—oh, my! Everything is complicated when you're working on a ceiling, craning your neck, reaching up above your head as you lean back to drive a nail. Everything is harder and guaranteed to cramp a score of new muscles. Try painting a ceiling for a hint of what it's like; you may decide to hire a professional.

The cost for having someone else install the finished ceiling ranges from wallboard (least expensive) to wood (most expensive). Costs for plaster and acoustic tile lie between these two, depending on the grade of material you buy. Wallboard is cheapest; the material is inexpensive, and skilled workers can apply it quickly. Plaster is also inexpensive, but it takes more time to apply. Acoustic tile takes time and, depending on the style you buy, the material can be very expensive. Wood is downright precious and the labor required to install it is considerable. Either wood or individual tiles are easy for the novice to install, though, making the cost comparable to other materials.

The lighting fixtures on the ceiling are just nailed or screwed there, with the wires running through the joists, to the wall, and down to the light switch.

The framing layer is, again, a bunch of edge boards, this time called *ceiling joists*. They are seldom as small as 2 by 4 inches but usually not larger than 2 by 8 inches. The ceiling is essentially an upside-down floor, supported by the tops of the wall (see Fig. 9-12). The joists are 16 inches to 2 feet apart; they are usually 2- by 6-inch or 2- by 8-inch boards if the house has another floor. As with the floor joists, the longer the span they have to bridge, the wider they must be. If interior walls can be arranged down the middle of a house 20 feet wide, for

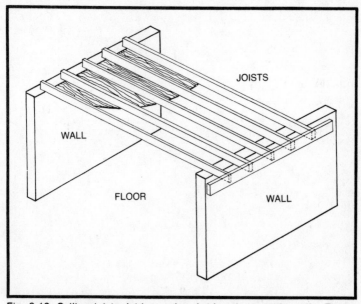

Fig. 9-12. Ceiling joists, laid on edge, bridge the outside walls. These boards are usually 2 by 8 inch thick. The tops of these joists are finished as a floor for the upper level, while the undersides are finished as a ceiling for the lower level.

167

instance, the span is only 10 feet—and 2- by 6-inch joists can be used.

If you didn't want interior walls at all, though, the unsupported span would be 20 feet; much larger joists, 2 by 12 or 2 by 14 inches, would be needed to support their weight without making the wallboard crack.

Plumbing

I gazed through the new window, watching a silent drizzle pelt the dusty ground outside, raising tiny clouds where the drops hit. The house was totally enclosed now. My mind said it was finished; but my loneliness clampled down worse than ever.

I started a fire but, disappointingly, still couldn't warm the house. I had nothing to do; I didn't have another wall to start, another window to install, another board to nail. I wasn't in the middle of anything. The plumbing or wiring was next, but they were clouds of dark, unknown, and forbidden territory in my mind.

I had a mental block against continuing. I went home—to our apartment—and took a long hot bath, trying to escape my thoughts. The house was built. Why couldn't I live in it now? Surely it couldn't be much longer before it would be done. Yet the plumbing would take a couple of weeks, as would the wiring, and so would many things after that. I wanted to fall asleep in the bath and wake up with the whole complex of difficulties washed away. It wouldn't work, though. I was in the middle of a long journey, and closing my eyes wouldn't stop the howling wind or the tumbleweed from brushing across the desolate state of my mind.

I sat there in the tub for a long time musing on the housebuilding events thus far, wondering how long it would be before I cut off a finger in a moment of inattention, or electrocuted myself. My plans for the future seemed remote. The house would take the rest of my life; I would simply die when it was over, either from exhaustion or old age. I didn't have the slightest desire to spend my life this way. I couldn't even enjoy the scenic beauty near the house in this state of anxiety. What was I doing building a house? People spent their lives working in these building trades. Who was I to think I could step in and duplicate a professional's work? Surely some catastrophe would make it all tumble down, even if I did finish it.

Little by little, I exhausted my charge of worries. Little by little, the dirty cream-colored apartment walls wandered into focus, securing me in their cramped closeness all around. The foggy mirrors, the dripping condensation on the faucets, and the single light bulb glaring from the ceiling brought me to attention. I stood up, dripping on the bathmat, ready for tomorrow.

PLUMBING DIAGRAMS

I studied plumbing for the next few days, using a 1933 library book and the "typical plumbing diagram" purchased with the original house plans. I assembled a rough diagram of the whole mess. I called the plumbing inspector who had earlier offered to help me, and showed him the spider-like diagram I had so painstakingly concocted. He glanced over it.

"Looks just fine," he said matter-of-factly. I was utterly amazed. "You gonna use *no-hub*?" he asked. I didn't understand. He rephrased his question. My old book had left me unaware of such plumbing advances.

"You mean there's a kind of pipe that doesn't need hot lead poured to seal the connections?" I asked.

"Sure. You buy *hubless* pipe and the special connectors for it. Just use a *no-hub wrench*, no lead." He also recommended using copper tubing for the supply pipe, since it would be maintenance-free, despite its initial higher cost. I agreed. I had vowed to never spend free time fixing the house or yard once it was complete.

Talking this lingo and holding my "approved" diagram was unreal. I still hadn't the slightest experience at plumbing.

Ignorant of the plumbing code, I asked him if anything in my diagram was forbidden by it. He looked more closely at the paper.

"Here and here," he circled several pipes. "Make sure you get seventeen-gage pipe." I memorized the fact, not understanding the difference, saying "okay" as if I did. Using that "typical" diagram faithfully, I included some superfluous parts. Hammer suppressors weren't necessary; he checked the half-dozen places where these showed up on the diagram.

"A good system won't hammer if you strap everything down good," he said. He spotted another code violation. "You need a cleanout here." He motioned to the diagram. "Looks good, Bill. Use cast no-hub and copper tubing, and you shouldn't have any trouble." He had more confidence in me than I had in myself, that was certain.

"I've been worried about soldering joints with copper tubing. Isn't that kind of tricky?" I asked. He assured me that it wasn't in the least, amused at my hesitation.

"What's the matter, Bill. You look a little piqued. Getting cold feet?" He was teasing. I didn't think it was funny, though. I was completely in the dark. He knew what was going to happen, but I didn't; and he couldn't take my anxiety seriously. He led me around the house, looking at where the pipes would go, pointing out a few shortcuts obvious only to his experienced eyes. "Don't worry, Bill," his mocking tone dissolving into sincerity, "you won't have any trouble."

I knew by know what I needed for piping or *rough* plumbing: 300 feet of several sizes of pipe and a bunch of bends and connections. Fixtures—sinks, toilets, bathtubs, and faucets—were something I knew nothing about. I must go shopping again.

The plumbing trade assigns a number to every part made. Each part has a name as well, usually something from Madison Avenue, like *Xandu* as the name of a bathroom sink, or *Desiree* as a bathtub. But devoted to efficiency, the sales clerk bothered only with the numbers. The plumbing materials, seductive sinks and provocative bathtubs, wound up as a sterile list of numbers on his order form.

I was certain that some magical trade secret was needed for this transformation to take place; I watched intently as he made up the order, starting with the kitchen sink Diane and I had already picked out. I showed it to him in the catalog.

"Okay," he said, consulting the table of numbers next to the sink photo. "That's a twenty-four-seventy-five-dash-sixty-three." He wrote 2475-63 on his sheet, then looked up. "Do you want that in any special color?"

"How much is it?" I asked. He took another large volume from below the counter and thumbed through it until he found the page he wanted. He ran his pencil down the column of numbers, stopping at 2475-63.

"Thirty-four ninety-seven," he said, "for white." He thumbed through a few more pages, again locating what he needed in a long column of numbers. All this time spent for a simple price; I learned that wholesale dealers don't have prices readily available. "Thirty-eight sixty-two for color," he said.

"White," I said.

"Do you want the P-trap and supplies?" he asked. Not knowing the plumbing slang, I didn't know what he meant. But I wasn't buying right now anyway. If he looked for it in the catalog, I could see a picture of it and see what it was.

"Ah, yeh, I guess so," I said.

"Trim?" I looked puzzled; he rephrased, his tone now slightly vexed. "Faucets?" He showed me some faucets—several dozen pages of them. I picked out the cheapest ones. We went through the other fixtures similarly until there was a two-page order. I only wanted a bid so I could shop around, so I asked him to figure out the price. Irritation rumbled in his voice.

"You'll have to come back this afternoon—say around four-thirty," he said. "It'll take me a while to price it out." I imagined him flipping tirelessly through the catalog pages for each of the items, and readily agreed to return.

I left jubilant; at last I'd have a price I could compare with other stores. I took a ride in the country and enjoyed myself until the afternoon. To a stranger, I suppose, it would seem like a soft life, being able to take time off like this—if my basic slavery to the job were ignored.

The next day I visited some other wholesale companies. Several places refused to do business with a noncontractor. But others were cooperative; I submitted a neat, professional list I'd copied from the one the clerk had made up.

"I'm building my own house," I'd say authoritatively. "I'm my own plumbing contractor. Will you price this order for me?"

They were happy to. They thought I knew what I was doing, with all those numbers neatly jotted on the page. And with all the numbers there, it took them no time to find the prices. If I had had a catalog, I could have made my own list. The prices now were 40% below those the first man quoted—40% seemed awfully high for just writing down some numbers. At last I had a genuine wholesale price! After a few days of thinking about these prices, I ordered the parts.

"Ol' buddy," said the clerk, "that no-hub stuff is a real blessing. When I was plumbing I spent a lot of time out there in the rain, pouring that old hot lead—splattered every time, got burns everywhere. You're lucky you don't have to do that, ol' buddy." It sounded pretty easy next to that, all right; my confidence rose.

"Are there any tricks, do you know, to making these copper connections with the blowtorch and solder?" I asked.

"No, not really. Well, yeh, you gotta make sure you heat the fitting, not the pipe, or else the pipe swells up and won't fit inside the fitting." Again, I simply memorized the hint, not knowing what he meant exactly. I ordered the first batch of parts, the no-hub pipe and copper tubing, leaving the fixtures for later.

Getting from the diagram and pile of raw materials to the first step wasn't so bad after all. Magic didn't exist; confidence was what I lacked. Only by going ahead with the plumbing, making mistakes and correcting them, could I finish the house. I attacked the problem cautiously, but with the attitude that everything could be redone if necessary. I transferred the diagram to the house itself, as I'd seen plumbers do in nearby construction, marking a line or an ✕ where each part would be. Then I picked up the drill and went to work cutting holes. The pipes went into place next, sometimes simply slipping in, other times fighting all the way.

Here and there I had to make bends around important house timbers that couldn't safely be cut through. I called the inspector when I wasn't sure if such bends or certain connections were allowed by the code. I traveled whole days, back and forth to plumbing stores and rental houses, looking for parts that would weave the ridgid pipe through the house frame as the diagram required. Professional plumbers have a distinct advantage over do-it-yourselfers—a truck always at hand with a supply of miscellaneous parts and tools.

The inspector said I had to test the piping right then, before the interior wall covering went on, to check for leaks. After several weeks of careful work, I expected all the joints to be watertight, but almost all of them leaked, to my dismay.

"Don't worry," he said. "Everybody has the same trouble." And later I observed another house where the professional plumber had similar trouble. Evidently, a plumber finds it quicker to race through the initial stages, knowing some joints will leak and some won't; those that don't will be quickly done, and the others will be taken care of after the test.

I became pretty confident by the time I neared the completion of the plumbing, particularly with the copper tubing. The joints in a few places were very close to wooden boards of the house and couldn't be connected except in position. At first I gingerly directed the torch in the direction of the pipe-and-wood combination, trying not to set the wood on fire. But after several hours of working on what should have been a five-minute operation, I got a bucket of water. Exasperated, I completed the joint with a normal torch flame, then flooded the resultant tiny fire with a bucketful of water, leaving the area damp for days to come. I repeated the technique thereafter as necessary, exalted at my mastery of the elements. I later found out that it would have been much simpler—and safer—to splash water on the beams *before* joining the connections. This way, there would be no small fire—which could quickly become a large fire—to put out. The beams would soak up the water and I could proceed to use the torch with little worry.

The pile of raw plumbing materials soon disappeared; I breathed a sigh of relief. My faith had brought me through; the treadmill vanished. The system worked; I hadn't had to study plumbing codes or materials or take classes!

PLUMBING SYSTEMS

For the sake of this description, I have grouped the parts discussed so far into two main systems: the *shelter*, which includes the walls, floor, and roof, and the *foundation*, which supports the shelter. The third main system is *mechanical*, the plumbing, wiring, and heating.

I used to think of these devices as nonremovable parts of the house, things growing out of the walls or floor, like little toes on the end of my feet. The sink was a baffling area where I could summon water from a pipe, then make it disappear again in the hole at the bottom of the bowl. Light switches were completely incomprehensible, turning night into day and back again.

When I installed the plumbing system I couldn't believe it would work. The inspector tested it; I tested it. But for a long time later I couldn't help thinking that something was missing, that it couldn't have been possible for me to simply stick a bunch of pipes together, copying diagrams from a book, and have a plumbing system just like one from my childhood visions. I hadn't taken any magic courses, I had only fastened a few pipes together. How could the system come to life without the magic that I lacked? At any rate, the plumbing worked. The mechanical systems are nothing more than a properly assembled bunch of parts.

The plumbing system has two main parts: the *rough* plumbing and the *finished* plumbing. What you see is the finished part, the sinks and faucets, the toilets, the bathtub, the washing machine, the dishwasher, etc. While each of these fixtures is different from the others in form and function, all have two things in common. They need something to let water

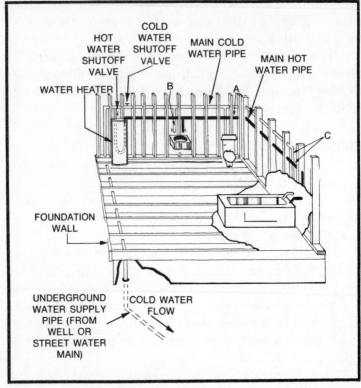

HOT WATER SHUTOFF VALVE

COLD WATER SHUTOFF VALVE

MAIN COLD WATER PIPE

MAIN HOT WATER PIPE

WATER HEATER

B

A

C

FOUNDATION WALL

UNDERGROUND WATER SUPPLY PIPE (FROM WELL OR STREET WATER MAIN)

COLD WATER FLOW

Fig. 10-1. A typical plumbing system, showing the supply-pipe routing through the house. The underground water pipe runs into the house from either a well or the water main in the street.

in, and they need something to let the water out after it has been used.

When you look at a sink, for instance, you see the hot and cold water faucets, somewhere near the bowl. Each one is fed by small pipes running inside the wall, one for hot water, one for cold. The wall is hollow between studs; the pipes fit between the studs before the wallboard goes on.

If the pipes are to run horizontally, holes must be drilled in the studs to run the pipe through (see Fig. 10-1). Cutting time can be saved by running the pipes parallel to and between studs or floor joists wherever possible. You can't see any of this, of course, unless you tear some wallboard down.

The pipes in the wall lead to the main water supply, which connects to the water well or public water main under the nearby street, as shown in Fig. 10-1. Only the cold water pipe goes all the way to the street, however. The hot water pipe goes only to the hot water heater, which is under the house, in a closet, in the basement, or almost anywhere (Fig. 10-1). The hot water heater is fed by its own cold water pipe. It heats the cold water and stores it in the tank. When you turn on the hot water at the sink, the pipe brings this heated water from the heater to the faucet. Some people buy extra-large hot water heaters (with bigger tanks) so they can use more hot water without running out.

Supply Pipes

This part of the rough plumbing, supplying hot and cold water to the faucets, is called the *supply*, and the pipe and fittings that go to it are *supply* pipes and *supply* fittings. The drainage part of the plumbing system uses larger pipe to let the debris through.

The sink is supplied with running water in the same way all the other fixtures are supplied. Sinks and tub/showers are the only fixtures with faucets, although washing machines, dishwashers, and toilets are no different. They have internal "faucets" not visible from the outside: *electromagnets* in the two machines and the *float ball* in the toilet. Look in back of the toilet, for example, and you'll see a pipe going into the fixture; it's a cold water supply pipe (toilets only need cold water).

The faucets are connected through a long pipe to the main supply pipe in the street. You may wonder how the pipe out there in the street knows when a faucet is turned. The water in the supply pipe is under pressure. If you poked a hole in the main water pipe, water would come squirting out—like in old movies when a car hits a fire hydrant. If you poked a hole in the main water pipe, then quickly jammed your house supply pipe into the opening, the water would go into the house instead of squirting all over the street. Pretty soon the water in the supply pipe would run through the faucet (if it were turned on), and *presto*, you have water in your sink.

This is what happens the first time you connect the supply pipe to the water main in the street. But no one wants to leave their faucets on all the time, so you turn it off, damming up the water. It seems as though you couldn't dam it up for very long; but, in fact, you can. Pipes are made to withstand the pressure of all this damming without breaking; and that's how a faucet works. The water is right there at the faucet all the time. dammed up, waiting to come out; all you do by turning on the faucet is open the dam.

Plumbing Design

The hardest part of plumbing is figuring out where to put the pipes in an economical way to conserve materials, and

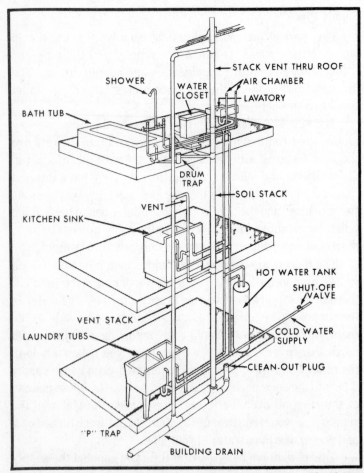

Fig. 10-2. The plumbing arrangement for a two-story house with a basement. Running all the fixtures from the same supply pipe will save time and money.

178

then putting them there. For a one-story house, tnis job isn't too difficult.

To design an economical system, you might first begin by running a cold water pipe directly from the main supply to each fixture, as well as a hot water pipe directly from the water heater to each fixture. The plumbing system would work okay, but giving each fixture its own set of pipes to the water source would cost a fortune and take forever to install. It would be ridiculous, but it's the worst that could happen to you, if you don't think at all.

The best that you can hope for is to figure out a way to run all the fixtures from the same supply pipe, as in Fig. 10-2, saving *money* by using the least amount of pipe possible, and saving *time* in the process because there are less pipes to install.

Figuring out the best system is pretty much a matter of trial and error for the beginner. An experienced plumber considers such things all the time and is paid for his speed. I started by running one pipe around to each fixture (on paper, of course) in turn. I found that by cutting over to a certain fixture I could use less pipe. Running the pipes under the floor (Fig. 10-3) lets you go directly from the toilet to the bathtub without wasting the pipe to go around the corner.

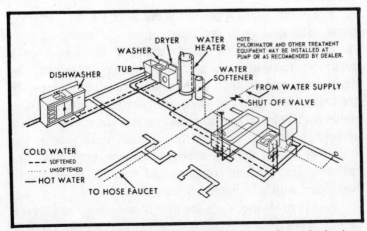

Fig. 10-3. A plumbing-fixture and water-supply layout for a single-story house. Running the pipes under the floor saves pipe that would otherwise be wasted by bending around corners.

The only thing to consider, besides the cost for piping, is the pressure at each fixture. You want a nice strong flow in the shower, for instance, not a trickle; and you don't want the bathtub to shut off when the toilet is flushed. To take care of this problem, the size of the pipe is what you have to think about. Three-quarter-inch-diameter pipe is the largest normally used in a two-bathroom house. If you ran the ¾-inch pipe to each fixture, you wouldn't expect any problem. But plumbers usually use ½-inch or smaller pipe where possible, since it's less expensive.

Over equal lengths of ½- and ¾-inch pipe, the ½-inch pipe provides more resistance to the water than does ¾-inch pipe, decreasing the pressure at the outlet. If you have, say, two toilets, a sink, and a shower in sequence supplied from a ¾-inch pipe, the shower would have better pressure than if you used a ½-inch pipe—especially while a sink or toilet is in use.

For example, in Fig. 10-1 the pipe from the street to the house (broken line) would most likely be an inch in diameter. From the end of the dotted section to the corner at A the pipe would be ¾ inch. The short pipes branching off this ¾-inch pipe (B) to the sink and toilet would be ½ inch. The pipes from the corner to the bathtub/shower (C) could be ½ inch if the runs weren't too long, and ¾-inch in diameter for optimum pressure.

Placing the pipes in position is the other job in plumbing a house. The goal is to put them where the plumbing plan requires with the minimum of cutting. If the house has a perimeter foundation, to protect the area underneath the house from cold weather, all horizontal runs are strapped to the underside of the floor joists, with vertical branches going up through holes in the flooring, between the studs, to their respective fixtures. In a house like mine, where the underside of the floor is exposed, the horizontal runs are between joists, when possible, and through them where necessary, with holes bored into each joist.

Also in my house, since the first-floor ceiling has exposed beams, pipes running between these joists for the second-floor water supply would be exposed as well. Rather than leaving them in view, where horizontal runs were needed to reach the second-floor fixtures, they were placed within the second floor walls, fitting through holes bored in the studs. With the studs only 16 inches apart, fitting a 5-foot length of pipe into the line of stud holes required cutting the 5-foot length into short sections that could be slipped between the studs and fitted back together when in place. Some people simply notch the studs for this purpose.

Kinds of Pipe

Several kinds of supply pipe exist, with different means of connection for each. Copper tubing, galvanized steel (zinc-coated to stop rust), and plastic are used in modern systems.

Copper is more expensive than steel, but it's far simpler to work with and requires only a propane torch and a hacksaw to make connections.

Steel pipe is threaded with a *pipe threader* (which is expensive even to rent), and requires more patience than working with copper. Plastic pipe is newer, cheaper, and connected with a special cement. Copper lasts forever; steel eventually rusts; plastic—who knows? I assembled the whole copper tubing system in two days, including all the connections. The novice can't go too far wrong on the supply piping. I had a few extra holes drilled needlessly and some leftover pipe, but everything worked out okay.

Fittings

Plumbing pipe is rigid and comes in straight pieces; routing it around corners and connecting it to other segments requires special fittings. The most basic and frequently used ones are shown in Fig. 10-4 with their names in *plumber's* language.

The most basic connection is the T-shaped intersection, called a sanitary tee, or *san tee*. One variation of it is a piece threaded inside (tapped) one or more of its branches to let a threaded pipe be attached, as at a sink connection. Another variation allows for pipes of different diameters to participate in the intersection: a reducing san tee.

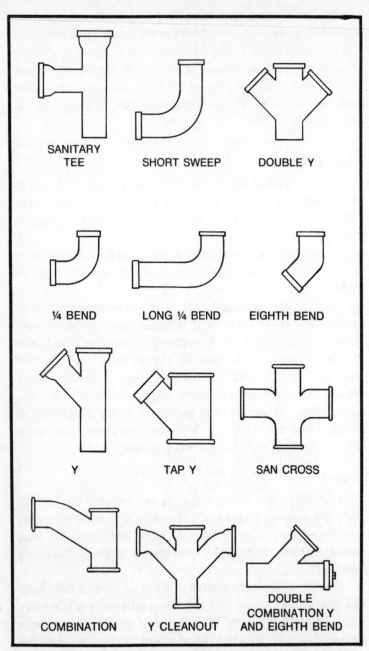

SANITARY TEE

SHORT SWEEP

DOUBLE Y

¼ BEND

LONG ¼ BEND

EIGHTH BEND

Y

TAP Y

SAN CROSS

COMBINATION

Y CLEANOUT

DOUBLE COMBINATION Y AND EIGHTH BEND

Fig. 10-4. Special fittings are required to route plumbing pipes through the various bends, turns, and corners throughout the house. The fittings shown here are those that are most commonly used, but there are hundreds of variations and combinations of these.

Other basic fittings are the *bend* and the Y (also called *wye*.) Various combinations of these basic parts are: the *wye* and *one-eighth* bend, the *double* Y, etc.

A *cleanout* is basically a wye with a threaded plug on one branch; once the house is built it allows you, after removing the plug, to insert a plumber's snake into the pipe to clean it out.

Hundreds more amazing combinations of loops, bends, multiple crosses, and wyes are available, but the ones shown here are the most common needed to plumb a house.

WASTE SYSTEMS

The other major part of the *rough* plumbing is the part that carries away the used water, the *waste* system. Returning to the earlier example of the sink, opening the drain lets the water into the drain pipe, which is hidden in the wall like the supply pipe. The waste pipe goes through the floor, still inside the wall, and under the house to a 4-inch pipe, called the *house sewer*. This large pipe either goes directly underground to the public sewer pipe in the street or goes to your own private septic system.

The waste system isn't under pressure, as is the supply system; it doesn't dam up the water. Only when used water is released from one of the fixtures does the waste pipe have any water in it (see Fig. 10-5). With the waste pipes empty most of the time, the sewer gases would be free to filter back into the house if it weren't for two precautions: *vents* and *traps*.

Traps and Vents

When I said each sink had a drain for letting water out, I omitted an important detail. Actually the water passes into the trap, and *then* into the drain pipe, as in Fig. 10-6.

The trap is the curved (J-shaped) pipe you see under the sink. It remains full of water, due to its shape, once water passes through it for the first time. Being full of water, it blocks gases from filtering into the sink. Every fixture with a drain has a trap like this, called a *P-trap* (Fig. 10-7). Toilets have them built in where you can't see them, as do washing machines.

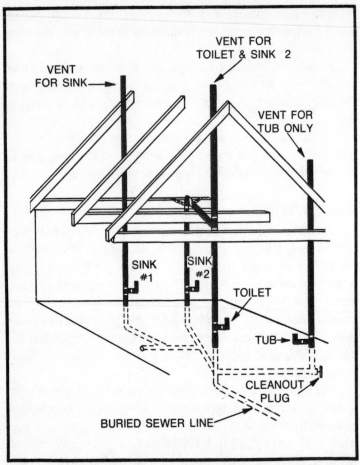

Fig. 10-5. A typical plumbing waste system. All pipes are hidden in the walls or under the house. The vents for sink number one and the tub are 1½ or 2 inches in diameter. The vent for sink number 2 and the toilet, called a soil stack, is 3 inches in diameter minimum.

It might occur to you that gas could accumulate behind the trap and somehow bubble through into the room when the pressure became great enough. The vent pipe provides another means of escape for the gases, however, eliminating this possibility. As shown in Fig. 10-6, the vent extends up inside the wall and goes through the roof. The vent exhausts the gases up above the roof, above everyone, harmlessly. Every fixture has to have its own vent, or has to share one with another fixture.

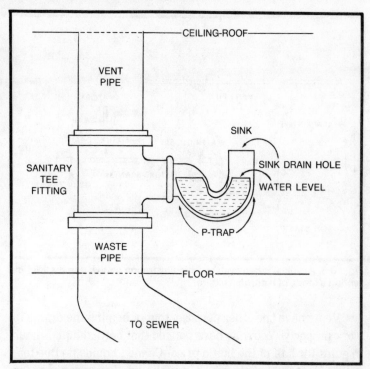

Fig. 10-6. The typical waste piping for a plumbing fixture. Water passes down the trap, then into the drain pipe to the sewer. The trap is filled with water to prevent sewer gases from leaking back into the house.

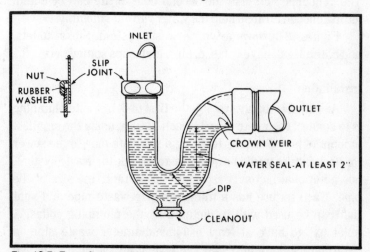

Fig. 10-7. Every fixture with a drain has a **P**-trap that blocks gases from filtering back into the sink. Toilets have them built in, as do washing machines and other equipment.

185

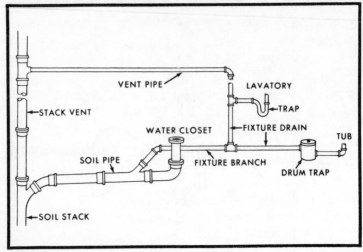

Fig. 10-8. A typical waste system. This illustration shows one method of venting a group of bathroom fixtures.

Vents have one other purpose, that of helping the drains to work properly. You may have noticed that bottled liquids pour in gulps for half of the bottle or so. Canned liquids do this too, unless a second hole is punched in the lid. The same goes for drains; liquid in a closed container needs air to pour smoothly. The vent pipe acts like the second hole in the can of liquid, letting enough air in to help the water run out smoothly.

Figure 10-8 shows a typical waste system. Sinks, toilets, and bathtubs all have vents, drains, and traps somewhere.

Installation

As with the supply system, the first task in waste plumbing is to connect the waste pipe for each fixture, using the smallest amount of pipe possible. Running a separate pipe to the sewer for each fixture would be quite wasteful. For the waste system, one additional factor is especially important; the size of the pipe. Each fixture has a minimum size waste pipe and vent that can be used with it, dictated by the plumbing codes. A toilet has to have at least a 3-inch-diameter waste pipe; a bathroom sink requires a 1½-inch-diameter pipe. Kitchen sinks, tubs, showers, and washing machines can use 2-inch waste pipes.

When several waste pipes come together, the code specifies minimum sizes. It might seem that two 2-inch pipes coming together would mean that a 4-inch pipe would have to be used from there on; not so. Few, if any, of these waste pipes are designed to be very full of water when in use. Rather than carry a lot of water, some have to be wide enough for large objects to pass through.

It turns out that (as any plumbing book will verify) a 2-inch pipe can accommodate several sinks and still be okay as far as the code is concerned. A 3-inch pipe can carry the waste from the whole house! It might not work if you flushed all the toilets and drained all of the fixtures at the same moment, but the chances of that occurring is nearly zero. A common waste system looks like Fig. 10-9. If you have several bathrooms, some of these sizes might change. A 4-inch pipe will carry two houses easily, perhaps three; no matter how many bathrooms you have, this diagram should be appropriate.

The ease of installing the waste system depends on the house. A conventional one-story house is the simplest to plumb; each fixture rests against a wall. The vent can go up inside the wall through the roof; the waste pipes and supply pipes can go down inside the wall through the floor, with everything connecting underneath as the diagram shows. The sewer goes out underground, where appearance doesn't matter.

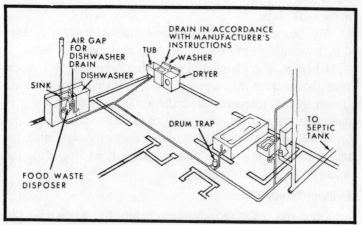

Fig. 10-9. The drainage system layout for a single-story house.

A house with a basement or second floor is more difficult; the waste pipes for fixtures on the first floor are easy to install, as if it were a single-story house, with the pipes running down the wall through the floor to the space below. The vent, however, has to reach the roof; and if the wall is an interior partition, chances are good that the second level won't have a partition over it for the pipe to hide in. If it doesn't, the vent will have to be led horizontally to a wall continuous with the roof. The plumbing codes limit the length of these horizontal vent runs, compounding the problem.

For second-floor fixtures, the vents present no problem; but the waste pipe does, since a hidden pathway must be found through the lower level to the space under the house. It may be necessary to make horizontal runs between the second floor joists, over to the top of a nearby first-floor wall. Neither case is impossible to deal with as long as you figure out how to get everything hidden before the house itself is begun; in case something won't fit, the design can be changed.

In my case, hiding the second-floor waste pipe was quite a trick, because the second-floor joists were exposed. Where the pipes could otherwise be hidden between floor joists under the ceiling board, this exposed design had no ceiling board; any pipes not inside the walls would be exposed. I solved the problem by running some pipes through the kitchen cabinets, since the second-floor bathroom was right over the kitchen. But it took very careful planning to make sure I got everything in the right place.

My house has few first-floor walls. Almost any other design would have more walls, making pipes easier to hide.

I had at first planned to make the necessary horizontal runs above rather than under the second floor, inside the wall. However, the inspector said the code wouldn't allow any horizontal waste pipe runs less than 42 inches above the floor level. Such a plan would only work for a sink anyway, since the toilets or tubs have their drain at floor level, requiring the waste pipe to run below that.

Drilling Through Studs

Once you've figured out where to hide the waste pipe, you need only cut the holes and put it in place. First go through the

house, marking where all the fixtures go, putting an ×
wherever the pipes are supposed to be. Then drill all the holes
in the floor (inside the walls) and through wall studs where
horizontal pipes are to go.

You might wonder if drilling a 3-inch hole in a two-by-four
stud doesn't weaken it. *It does.* Walls where waste pipe will be
hidden are normally made with two-by-sixes or two-by-eights,
so that after you drill them to pieces they still have some wood
left to hold up the ceiling. Also, arranging pipes inside the wall
is easier with more space. A 3-inch pipe will fit in the hollow
space of a wall with two-by-fours for studs; but fittings, bends,
junctions, and even straight couplings, are larger than 3 inches
and simply won't fit in the space.

It's sometimes necessary to have 2-inch pipes cross each
other inside the wall. This would be impossible in the 3½
inches of space afforded by two-by-four studs. But two-by-six
studs leave a space of 5½ inches and would work nicely. I was
very careful to bend the pipes around top and bottom plates,
where they went through floors in load-bearing walls, since
cutting a 3¼-inch hole in a two-by-four is just like cutting it in
half. Since the plate of such a wall keeps the studs and the wall
itself in line, cutting it would seriously weaken the wall,
making it vulnerable to high winds or other forces. Figure 10-5
is shown with extra-thick walls.

I managed to hide the few bulges I had behind cabinets or
in closets. I not only wasted a lot of time figuring out how to
hide these blunders, but I spent a lot of emotion beforehand
worrying whether or not I could do so. On top of that, I had a
difficult time doing the drilling and fitting the pipes in place.
It's much easier to make the walls thicker in the first place,
saving yourself time and trouble; no plumber would ever
bother trying to hide pipes in a two-by-four studded wall.

Special Framing

Another point related to the house itself is the special
floor-joist framing needed for toilets, tubs, and showers.
Unlike sinks, attached to a waste system in the wall, these
fixtures sit directly on the floor. If it happens that the fixture is
located directly over a floor joist, you can't just cut the joist to

connect the waste pipe in the correct position. You have to install a *header* joist across the two adjacent joists. This gives you a place to cut the hole you need.

Waste-Pipe Materials

Two materials available for use as waste pipe are *cast iron* and *plastic*. The plastic type costs less, is easier to work with, and is lighter. You can cut it easily with a hacksaw and make connections with glue. If you drill a line of holes in the studwork, and they don't quite line up with each other, you can bend the pipe slightly to get it through the holes by heating it with a torch.

I've been told that a person can install the waste pipe in an average three-bedroom, two-bathroom house in two days using plastic pipe. The only drawbacks are that it's against the building codes in some areas, and it's said to be noisy.

Cast iron is hard to work with. (It took me two weeks to install it in my house.) It's very heavy—not hard to lift, just tiring. You need a special expensive tool to cut it, which I rented for $5 a day. The modern cast iron pipe is called *no-hub*, and it's joined with stainless steel and rubber compression bands, which heed a special *no-hub* wrench to tighten properly. The wrench cost me $11. The connectors cost 80¢ to $1 each. The pipe cost me about $1 per foot.

One thing remains regarding waste pipe in horizontal positions: In all cases it must slope toward the sewer with at least a quarter-inch drop for each foot of pipe (Fig. 10-5). The slope can be as great as desired, as long as it meets the minimum requirement. In rare cases, the sewer isn't low enough for the house sewer to have the required slope on its way out, a situation common on hillsides. In this case, an electric pump must be installed.

Plumbing Tricks

Put the house water shutoff valve in a *convenient* location.

Be sure to *nail* the outdoor hose faucets in place after they're connected; otherwise you may pull them out of the walls.

Install sink faucets and drains on each sink *before* the sink goes on the wall or in the countertop.

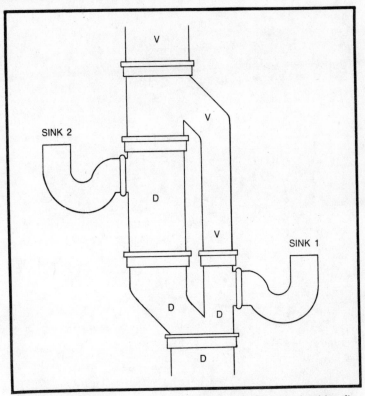

Fig. 10-10. A special trick for using a single drain for two plumbing fixtures, sitting one over the other on different floors, is to use a single drain and vent pipe. In this illustration, D is for drain and V is for vent.

A one-piece fiberglass shower/tub is too big to fit through most doorways or halls; place it in the bathroom area before the walls are framed in.

Copper tubing connections require the ends of the pipe to be cleaned with steel wool and coated with soldering flux before the solder is applied. Cast iron pipe is easily cut with a special cutter rented from any tool-rental place. Cast iron sections don't have to line up perfectly at connections; the rubber connector will bend a little without functional difficulties.

A very specific trick for installing two plumbing fixtures, one above the other on different floors, is to use a single drain pipe and vent pipe as shown in Fig. 10-10; D is for drain pipe (waste), V is for vent.

Electricity  11

Finally came the wiring. As if things weren't difficult enough, I could see that the money would soon run out. And as I started reading about the wiring, wondering how I could talk a bank into loaning me money, I was once again depressed at having to leave an area of proficiency and plunge into a new unknown.

It's hard to recollect this period of time; no fond memories, no concrete achievements marked this phase of building. I visited my old housebuilder friend, but he was too busy now; he wanted me to figure it out from scratch like he'd done, perhaps, and evaded my questions.

"I'm having a little trouble with planning the circuits," I said. "Can you help me?" He was on a ladder, installing wood paneling. I squeezed the comment between saw bursts. The saw went back on, sawdust filtered down on me.

"Oh, there's not much to it," he said, when the saw stopped again. "You using number-twelve wire?"

"Yes."

"Well, nothing to worry about. Just put ten or so lights or outlets on each circuit. That should work out okay."

But it wasn't that I was worried; I simply didn't know what I was doing! I didn't even know the objective of wiring at that point. I could have been studying how to string rope between outlets, instead of wire in a parallel circuit, for all the understanding I had of wiring then. It was looking as though I would have to make another great leap, propelled by faith alone.

"The book mentioned something about calculating the number of *watts per square foot* in the house, or counting the *current load*. What about all this *load* business in the book?" I said. "Don't you have to know the watts or current for each fixture?" He thought it funny that I was using a book.

"Oh, nah, I didn't use any books. Just common sense pretty much." he said.

"Well, didn't you have to figure out watts or something for your circuits?" I asked.

"Well, no. I'll tell ya, all you need to do is just put ten or twelve devices on each circuit. That'll take care of things."

I simply couldn't trust him with that reference to common sense. I didn't have any if that was true. The book made it seem so complicated. He made it seem ridiculously simple. Surely, either he or the book was wrong. Even if he'd used his method on his own house, who could tell whether or not it would burn down or electrocute someone. I was inclined to trust the book. Time was wasting. Maybe if I slacked off for a few days, I'd get a new insight to the problem. Money was a pressing worry. I could work on it now.

MATERIALS

One thing I'd picked up from my taciturn common-sense friend was that wiring was supposedly very easy. Blind faith had conquered the magic in plumbing; I again put my trust in the unknown and my friend's advice, and made a diagram assuming everything I knew was right, that nothing was more complicated than what he said. I assumed the book that frightened me had been written poorly—following the simplistic advice of my friend would work.

I started from scratch. The house had wires. It had a service box where the "big" wire from outside connected to the "little" wires inside. Each wire leading from the box was called a *circuit*. Each wire inside either went to a major appliance, oven, water heater, furnace, or dryer (I have an all-electric house, including furnace), or it went to a string of outlets and light switches, not exceeding 10 in number. I figured out where the wires would be hidden in the walls or floor, and drew up the diagram illustrating this placement. I showed it to the electrical inspector, hesitantly, expecting him to laugh at everything I'd done wrong. But he said it looked fine! He recommended several changes, but other than that I wound up with a plan for the wiring.

"You have to use metal utility boxes, now," he said. "Your friend down the street used fiberglass and he had to tear them out. Do you know how to figure the sizes?"

He told me that the boxes the outlets sit in must provide a specific amount of space for the connections inside to be made.

"I'll show you how to ground the service box." He indicated on the diagram where a ground wire had to run from a water pipe to the neutral bar, inside the service box. "You have to ground all utility boxes and the devices inside them with the bare (fourth) wire inside the Lumex cable. The oven can be grounded to the neutral (third) wire in the three-conductor cable, though."

I memorized these words, not yet knowing what they meant. We chatted for a few minutes; but as soon as he left I plunged into anxiety anew. I didn't know anything about what kind of wire I had to use; I didn't know anything about any of the parts, or what any of them even looked like. But I couldn't waste a moment more learning anything unnecessary; I had to learn the rest of wiring as I went along, as I did with the plumbing.

I knew that the parts would be standardized; any good salesman could get me the right parts, since they would be very common, if I simply told him the name of the part. If I could just find a helpful salesman and tell him I wanted the cheapest models, I wouldn't need to know what anything was, since he'd get them from a storeroom himself anyway.

Once I had the parts, I'd call the inspector and he could pick out any noncode pieces. He'd tell me the names of the preferred parts, and I could exchange the rejected ones for the right parts. I'd have all the correct parts without having to first learn one from another; as I went along, I'd get to know the different ones.

I started out according to this plan, ignorant. In hopes of avoiding the run-around I had with plumbing materials, I took the advice of my common-sense friend and went directly to the electric supply company he'd dealt with. "Just ask for Ronny." he said. "He'll fix you up."

"I'm a contractor," I said to the woman who offered her assistance, "and I need some light fixtures." I surveyed the walls and ceiling, both completely covered with every imaginable variety of fixture, each trailing a tag below it. I reached for the one nearest me. The numbers on it were undecipherable. "How do you read the prices?" I asked. "Or aren't these prices?" I thought perhaps, she had a price list she'd have to consult before telling me any prices.

"Oh, just read the bottom line and multiply by forty percent—you'll get the normal contractor's discount, sixty percent off." It sounded like a bargain. I found one I liked.

"How much is this one, then." I held up the tag to her.

"Forty percent of one-hundred and twenty-five dollars—fifty dollars." She had a clipboard with a sheet of paper, listing the different rooms of a house on one side. "Would you like that one?" she asked. I hadn't come to buy, but to compare. I might as well find out what a few things cost.

"Yes, ah..." I looked at the list on her paper. "...for the dining room." She wrote down the manufacturer's number and name of the light. The pencil lead broke, and she went off for a moment. I spotted another fixture for the kitchen. Before I could read the tag myself, she grabbed it.

"Ten-fifty, for the contractor's price," she said. "Here's one for only five-fifty." She pointed. I liked it okay except for an ornate globe.

"Can you get a simpler globe for that—just a plain sphere, perhaps?" I asked.

"Oh, I'm sure we can," she said. "Let me check with Ronny, if he's back yet. He'll know." In a moment she returned. The light was available as I wanted it. We strolled around the room, picking out other fixtures until her sheet was filled with writing. Buying like this wasn't much fun at all; it was pure work, trying to get something economical yet pleasant looking.

"Shall we put that on will-call for you, then?" she said. I thought she was simply helping me price things. I hadn't thought she viewed this as an order. I couldn't order here without first comparing prices elsewhere. I never bought things on first inspection anyway. I hoped she wasn't angry.

"Well, why don't you just hold onto the order for now. I'll have to check into my finances before finalizing it." She agreed, smiling. I still needed to see Ronny, I said. She gave me her card and led me down some passageways, through several doors, and into a room with many desks. At one was Ronny. I told him I needed some parts for the wiring.

"Well, your electrician'll worry about that for you. Who's he?"

"I'm the electrician," I said defensively.

"Oh, I see, I see!" he said. "I'm sorry. So you're doing the wiring yourself, eh?" He looked at me out of the corner of his eye, in a lopsided, warm way. I nodded and said I was.

"Well, let's see now. What do you need? Do you have a list?" I showed him the list of parts I thought I needed. "Oh good," he said, seeing the list. "Come on over here and we'll make out an order for you." We went out to the display room to a counter. He rustled around underneath and dredged up an order book, arranged some carbons, found a pencil, and settled with his elbows on the counter. "Okay, what's first?" he said in a helpful, fatherly voice. The moment of truth was at hand; I consulted my list.

"Fifteen light switches," I said. He paused.

"What kind of wire you using, Bill?" he said. (I shrugged.) "Lumex? Twelve-two? Twelve-three? Fourteen?" I recognized the name "old common sense" had mentioned.

"What's the twelve-two or twelve-three business mean?" I said.

"Are you in the city?"

"Yes, my house is." "Well, you need twelve-three then for the code. You have three wires in twelve-three," he said, "one hot, one cold, and one ground, all *twelve*-gage wire." He wrote down a number for the light switches. "How much wire do you need?" My plan was working!

"Eleven-hundred feet," I said.

"Okay, that comes in two-hundred and fifty-foot rolls. Do you want four or five rolls?"

"Four," I said, thinking that perhaps I had overmeasured. I could always buy more.

"What else now?"

"Eighteen outlets," I said.

"Okay," he spoke as he wrote, "eighteen duplex receptacles. Brown or white?"

"White," I said. We went down the list: electric meter base, staples, utility boxes, light-fixture hangers, wire connectors. He put down the right numbers each time.

"What kind of wire do I need for an electric dryer?" I asked.

"That would be number-six Lumex," he said.

"And I need a service box and circuit breakers," I said.

"How many circuits are you gonna have, Bill?" I didn't know exactly. I showed him my diagram. He counted them. "Okay." He thumbed through a catalog of boxes and circuit breakers, stopping at one. "That should be a 5 × 4592." He wrote down the number. "And a circuit breaker for each circuit." He looked through the diagram at each circuit, marking down a switch for each.

He gave me a copy of the list, but without any prices. I asked him to price them, since I wanted to compare.

"Oh, don't worry, Bill," he said. "I'll figure this out over the weekend." (It was Friday afternoon.) "I'll have it for you Monday."

"Okay, go ahead and make up the order," I said. I couldn't wait any longer for the parts. He'd helped a lot, making my instant-learning plan work. I wanted to start wiring. He said it would be ready Monday afternoon.

At last I carried my new parts to the house. Electric parts had always intrigued me; they didn't threaten as the lumber and plumbing supplies had.

I called the inspector; almost everything was proper. I exchanged a few wrong parts and started working. Dealing with the parts only thus far had already taught me most of what I needed to know about them and the ways of installing them

I nailed switches, outlets, and light-fixture boxes in place in the rough framework, weaved the wires through the holes I'd made for this purpose in the studwork, as planned in the diagram, and made the connections in each place. I nailed the service box into place, installed the circuit breakers, and connected the wires inside to the circuit breakers. (I visited a display at the electric power company to see how to wire the circuit box.) In 10 days the wiring was complete, except for the furnace.

FUNDAMENTALS OF ELECTRICITY

Electricity is quite a bit more complicated than plumbing. I started the house with some experience working with basic theory, but ignorant of house wiring. I had a healthy respect for electricity, though, knowing there wasn't any visible difference between a harmless wire and one with enough power to electrocute me. Of course, no power was connected until the wiring was thoroughly inspected by the authorities, making me feel safer.

A lot of household appliances have to be "plugged in" to work; lamps, televisions, electric typewriters—all work on electricity. The outlet has to get the electricity from somewhere, just as the faucet must be attached to a supply pipe. Some wires lead from the outlet through the walls (again like the plumbing) over to the load center, also called the circuit breaker box, fuse box, or the "service" (see Fig. 11-1).

The load center connects to the electric meter that the power company uses to see how much electricity you've used. A wire runs from the meter to the power line in the street, usually a wire on a power pole (Fig. 11-2). Each switch, light fixture, and electric appliance must connect to the load center in a similar way. But like the plumbing fixtures, if each outlet, switch, and fixture had its own set of wires running to the load center, a lot of wire would be wasted, since one wire can serve several devices.

In actual house wiring, however, a single wire doesn't run from each outlet to the load center, but from one outlet to the next, to a total of eight or ten devices in a row before finally leading to the load center. The load center winds up looking

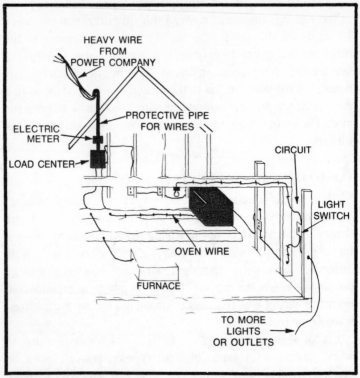

Fig. 11-1. A sample electrical system for the home. The load center is located within the house. It contains the circuit breaker or fuse box. The electric meter is outside the house. Notice the typical outlet circuit stringing through holes in the studs (predrilled before installation), going from outlet to outlet. Also notice that "heavy" appliances, such as the furnace and oven, have their own circuit.

like a big spider, because an average house has several of these strings of outlets and light switches wired in *parallel*, supplying all its outlets and light fixtures.

Additionally, each major appliance (such as the oven and the furnace) has its own line directly to the load center. Most houses have several major appliance lines (circuits) and several more strings of outlets, all coming into the load center. The load center itself is connected by a single set of wires to the meter, and the meter connects with a single set of wires to the power company's cable in the street.

Some of the basics from Fig. 11-1 are as follows: Each outlet, switch, and light fixture, and all wire connections are

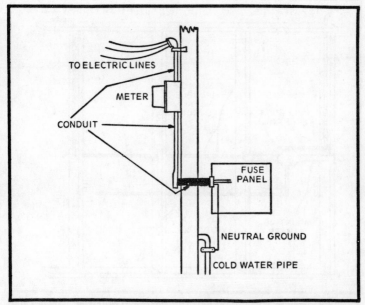

Fig. 11-2. An above-ground electric service installation. The power comes to the fuse panel (or circuit-breaker panel) from the electric power lines, through the conduit and service meter. The fuse panel is grounded, as are all individual circuits, to the cold water pipe.

required to be in metal *utility boxes* (Fig. 11-3). Depending on how many wires and switches are to be connected, the size of the box will vary; the electrical code strictly specifies these sizes. The inspector cares a lot about the sizes being correct, because the wires mustn't crowd each other where connections must be made; ask him first if you're not sure.

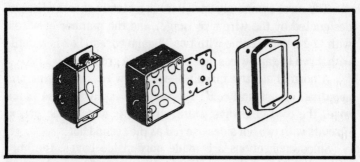

Fig. 11-3. Some common types of outlet or utility boxes. Each outlet, switch, and fixture must have its wire connections made in one of these metal boxes.

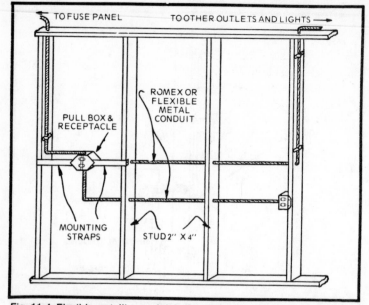

Fig. 11-4. Flexible metallic conduit is the easiest to work with. The flexible conduit is strung through holes that are predrilled in the studs before they are nailed together to form the framework. Sometimes the flexible conduit can be run in "notches" on the room side of the studs, rather than holes drilled into them. This method is especially applicable for rewiring older homes.

The electrical system consists of all these devices: outlets, switches, light fixtures, and appliance outlets, and the wires between them. Between boxes you can run the wire through the walls, drilling through studs (Fig. 11-4) as necessary, and stapling the wire in place, with a staple every four feet as the code requires.

The most commonly used wire is a plastic-sheathed cable, designated by the wire size (gage) and the number of wires, with 12-2 being a cable with two 12-gage wires; 12-3 is a cable with three 12-gage wires, one of which is a ground.

A normal electric circuit needs two wires, a *positive* and *negative*, or *hot* and *cold*. The hot side is always the *black* wire; the cold is always *white*; ground is always the *green*. Circuits with two hot sides use *red* as the second hot.

Since connections are made only inside boxes, the only wiring outside of these junction centers is the single-piece cable; the wires inside may number 2, 3, or 4.

WIRING

. Wiring consists of two steps: *calculating* which wires should go where; and *connecting* them in place. At first it might seem simpler to buy a bunch of wire and run it around the house to all the fixtures, outlets, and switches in turn. But were you to do just that, the wire would most likely be forced to carry more current than it could safely and would overheat, possibly burning down the house.

The goal of electrical wiring is to evenly distribute the massive amount of electricity coming from the street to the load center, so that no wires are ever forced to carry more current than they safely can without overheating.

The simplest way to achieve this goal is to figure out how much current will be needed in each cirucit, and to install wire large enough to carry the current flow without overheating. Electrical code tables dictate what size wire can be used safely, without danger of heating, for any given current. Table 11-1 lists common wire sizes and their maximum ampere ratings. The wire sizes are according to American Wire Gage (AWG) standards. The ratings are for copper wire, either solid or stranded. Note that the capacity increases as the wire number decreases.

Once the wire size is chosen, make sure no greater current than it's ratings will ever flow in a specific set of wires. You achieve this by installing an overload device between the load center and each circuit (a circuit is a complete two-wire path).

Table 11-1. Wire Size vs Ampere Capacity

WIRE SIZE NUMBER	AMPERES
18	10
16	15
14	20
12	30
10	40
8	55
6	80
4	105
2	140
0	195
00	225

Table 11-2. Load Limits in Watts for Various Capacity Circuits

CURRENT CAPACITY, AMPERES	LOAD CAPACITY, WATTS
15	1800
20	2400
30	3600

An overload device can be a circuit breaker or a fuse; its job is to open the circuit, shutting off the electricity whenever too much current (amperage) goes through the wires. Whenever too much current flows, the circuit breaker or fuse stops the power before the wire has time to heat, melt its insulation, and cause trouble. Table 11-2 gives the maximum load limits for various circuits. You can add the wattage of the lights and appliances on a circuit to determine the *total* circuit load. Fuses melt at the rated current flow, breaking the circuit; circuit breakers simply switch *off*; you don't have to replace them, just switch then back *on*.

For a string of outlets, switches, and light fixtures, the usual expected maximum current is about 20 amperes (20A). Number 12 wire, according to official tables, will carry 20A okay; so, to make a circuit running to a bunch of general lighting devices (switches, outlets, and fixtures), 12-3 sheathed cable can be used, with a 20A circuit breaker or fuse protecting it at the load center.

According to Ohm's law, a basic electrical rule, voltage multiplied by current equals power, measured in *watts*. If you had a 20A circuit with the standard 120V power source, the maximum wattage that the circuit could supply would be 20A times 120V, or 2400 watts (2400W).

If you turned on twenty-five 100W bulbs at the same time, this circuit would shut itself off. For 10 or 12 devices on a single 20A circuit, the chances of using all 2400W isn't very great, since having two 100W bulbs turned on at every outlet isn't normal. Yet several 100W bulbs are probably turned on at the same time each day; add to them a television at 300W, a stereo at 100W, and an electric heater with 1000W, and the circuit gets a workout. Table 11-3 shows sample wattage values for various devices.

Kitchens usually have their own 20A circuits with only one or two outlets, since a toaster can use 1150W all by itself; turning on a 1150W toaster and a 2000W frying pan would open the circuit, most annoyingly, in the middle of breakfast.

Major appliances need more power, sometimes 5000W for an oven/stove, so much heavier wire is used. Number 6 wire, for example, might be employed with a 40–50A fuse or circuit breaker on a single circuit.

In Fig. 11-1 the wires leading to the outlets, switch, and light fixtures are number 12; the wire going to the furnace is larger, say number 6 for a 7000W furnace, and to the oven number 6 again, assuming it needs 5000W at peak loads. The wires leading out of the house to the power lines are enormous: 00 (pronounced two-ought) can carry 200A without heating. Wire gage numbers go down as wire size goes up: 1 is smaller than 0; 00 is larger than 0.

Table 11-3. Average Wattage Values for Various Consumer Devices

APPLIANCE	POWER, WATTS
Air conditioner (window type)	1560
Blender	250
Broiler	1430
Clock	2
Clothes dryer	4850
Coffee maker	895
Dishwasher	1200
Electric frying pan	2000
Electric range	12 200
Garbage disposal	900
Hot water heater	2500
Iron (hand type)	1080
Light bulb (100W)	100
Radio	70
Refrigerator/freezer (14 cu ft)	325
Refrigerator/freezer (frostless)	615
Television (black-and-white)	240
Television (color)	330
Toaster	1150
Vacuum cleaner	630
Washing machine	500
Window fan	200

The power company complicates things a little for the beginning electrician by supplying the average house with three wires—two *hot* and one *cold*. The second hot is to provide 240V service.

The net result is that you can supply an appliance with a 240V power source by running the two "hot" conductors to the appropriate terminals, and the white (neutral) to the grounding terminal along with the fourth, noninsulated ground wire.

Most large heating elements, such as those in a furnace, dryer, oven, or water heater, are more durable when they use 240V service rather than the standard 120V. Every 240V circuit is simply two 120V circuits stuck together, with a single double-action circuit breaker. Red is the other hot in 240V circuits. The 220V circuit has three wires—black, red, and white in addition to the bare ground wire in each case.

I might not have done the wiring had Lumex (flexible plastic cable) been prohibited. Without it, metal tubing (conduit) would have been the alternative; but it is like plumbing pipe to install and quite expensive (Fig. 11-5). The Lumex was simple to put in; I threaded it like rope through the holes drilled in the studs. It only took a single day for this drilling and threading. Making the connections took another day.

Each 120V circuit has a Lumex end coming into the box, with the *hot* (black) wire going to the circuit breaker and the *cold* (white) wire going to the neutral bar, as does the bare metal ground wire. The sole function of the bare ground wire is to connect all utility boxes electrically to the neutral. If the hot wire in any box ever touches metal, a short will occur and the breakers will turn off the power before anyone has a chance to touch the "hot" box and get shocked or electrocuted.

Each 240V circuit has a Lumex end with two *hot* (a red and a black) wires and a *cold* (white) wire coming into the box. The two hots go to the circuit breaker (240V circuit breakers have openings for two wires), and the cold goes to the neutral bar, as does the bare ground wire.

Few tools are needed for wiring: wirecutters, a hacksaw (for heavy cable), pliers, a drill, and a screwdriver. The drill,

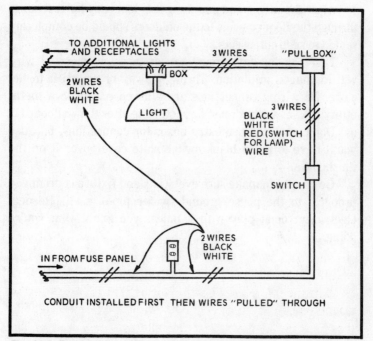

TO ADDITIONAL LIGHTS
AND RECEPTACLES
3 WIRES
"PULL BOX"

2 WIRES
BLACK
WHITE

BOX

LIGHT

3 WIRES
BLACK
WHITE
RED (SWITCH
FOR LAMP)
WIRE

SWITCH

IN FROM FUSE PANEL

2 WIRES
BLACK
WHITE

CONDUIT INSTALLED FIRST THEN WIRES "PULLED" THROUGH

Fig. 11-5. Rigid cable entails a much more complicated installation procedure. The conduit is installed first, then the wires are pulled through it.

with a bit in place, must either be small enough to fit between studs or have a right-angle drive to fit in; I spent two whole days, believe it or not, searching rental yards for such a tool.

WIRING HINTS

When placing light switches for a room, note which way the door swings so you can avoid putting the light switch on the hinge side.

Plan to give the living room two separate circuits, so someone can run a hair dryer without creating static in the television circuit.

The telephone company often prewires houses for phones in every room, should someone in the future want them; they thread the wires through for free when the house is in the framing stage.

Wires for a house-wide stereo system could be installed during the wiring stage, too.

After the wiring and plumbing are done, any holes leading through the floor or walls to the outdoors should be completely sealed to keep out insects.

Don't, under any circumstances, use any piece of wire with damaged insulation. Throw it away or return it to the store. No wiring connections are allowed *except* in a metal utility box; every box must be accessible after the wallboard is on. Don't, as I did, use extra boxes for connections, because you'll have to put a blank switch-plate cover over it on the finished wall.

Be careful to make sure every box and fixture is grounded properly to the house ground (water pipe) itself. Respect electricity; don't play with it unless you know what you're doing.

Heating the Home

I originally planned to use electric baseboard heaters in each room. Then an acquaintance mentioned that he worked for a company that made electric furnaces and could obtain one for me wholesale. I liked the idea of having a furnace without needing a chimney for it. I was too low on ambition by then to build a chimney if I could avoid it.

I was afraid to take him up on it, though, thinking that the *ductwork* (piping that carries hot air from the furnace to each room) was expensive and hard to install. Another acquaintance, though, said he had once worked installing ductwork, and that it was simple. By now the blind faith aspect didn't bother me; I looked forward to conquering another new area without previous experience. I took them both up on it and bought the furnace.

The first friend drew up a diagram of what size pipes to use in the ductwork, and where to put the hot air registers for each room, as well as where the air intake ducts should be—all in 15 minutes or so, since he knew all the technical requirements so well. He sent me to another ductwork company for the sheet-metal pipes, where I was greeted as a representative of Mr. William Heating Co. I picked up all the metalwork at one time, but made a number of errors in buying and had to return twice a day for the rest of the week. The two guys who filled my orders were hostile, at first, to my apparent inexperience. They became friendlier and friendlier as the week went on, however, as I returned again and again.

"Boy, you've really got a lot of business, it looks like—with all this stuff you're buying," said one, not realizing everything was in exchange for something else. Just seeing me so often, though, had given the illusion that my business was booming. Appearance is indeed a powerful tool in the business world!

With the furnace connected, the wiring was complete. The inspector inspected. The power company inspected. When all was approved, a crew arrived to connect the electrical system to the power line.

A new plateau of progress had arrived. I had pulled myself up to an understanding of house wiring by my bootstraps; I had made a house wiring installation without prior experience, by drawing up a diagram, having it criticized, finding all the correct parts, and finally putting it all together—with blind faith in myself and my advisers as the guiding force. Wiring had indeed been fairly simple. The worst part was running back and forth to the electric store for 10¢ parts that an electrician would have carried with him.

HEATING SYSTEMS

Quite a few different kinds of heating systems are possible, falling into two main categories: *radiant* and *forced air*. If you have a furnace, it's forced air, since the air is heated in one place and carried through pipes or ducts to where you want it. Radiant heating is the kind where a heater warms the room directly, without blowing air around.

I originally planned to use *baseboard heaters*: long, thin boxes that sit on the floor along a wall, wired into the house wiring system just like a water heater or oven, by a long element in the box. Since they get hot right in the room, you can't safely have floor-length drapes without creating a fire hazard; small children might stick their hands in and get burned, too. Some in-the-room radiant heaters are built to use hot water or steam, rather than electricity. But these are plumbing projects all by themselves. As a novice, I wasn't interested in trying them.

Some people put wires in the floor or ceiling, acting as an electric blanket for the room. Some run hot water pipes through the walls and floor to act similarly; not only does this cost four times as much as electricity, however; it's harder to install.

With these drawbacks in mind, I decided that forced-air heating would be best. With forced air, the only evidence of the heater in any room are the openings in the floor or low on the wall (Fig. 12-1). The air is heated in a furnace and carried to these openings by large sheet-metal pipes (ducts), which are easy to install. The furnace is usually located near the center of the house, either in a closet or below the floor, using the least amount of ductwork to carry the hot air from the furnace to the rooms; hence the name *central* heating.

There is a furnace available to use each type of fuel; coal, oil, natural gas, propane, or electricity. Whatever you decide on, the part that burns the fuel and heats the air is bought all in one piece at a heating supply company; the ductwork is purchased in pieces and assembled at the house.

To install the heating system, all you do is calculate the size of furnace you need, buy the furnace and ductwork

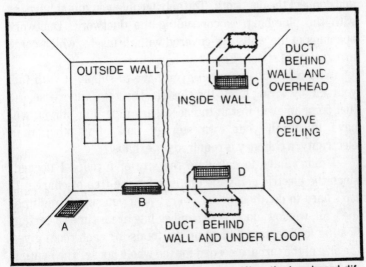

Fig. 12-1. For an outside wall the floor register (A) or the baseboard diffuser (B) can be used. Notice that the floor register is a few inches away from the wall, while the baseboard register is against the wall. For an inside wall, the high wall register (C) or the low wall register (D) can be used.

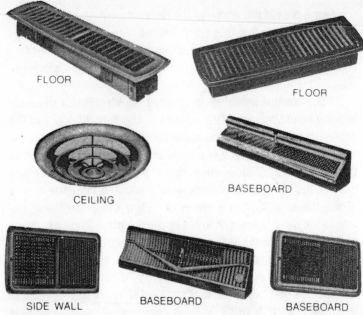

FLOOR

FLOOR

CEILING

BASEBOARD

SIDE WALL

BASEBOARD

BASEBOARD

Fig. 12-2. Diffusers for various applications.

sections, cut holes in the floor where you want the hot air to come in, set the furnace where you want it, supply it with fuel, and connect the ductwork. The only trouble you might have, as with the plumbing, is concealing the ductwork. Ductwork openings in the walls are covered with *diffusers*, which come in a variety of styles (Fig. 12-2).

With an electric furnace, you have to supply it with fuel yourself, by wiring it into the electrical system. Otherwise, the fuel company will install the necessary pipe. For oil you will have to install your own storage tank. For other than electricity, a chimney is required for exhaust.

I didn't have to calculate the size of furnace I needed, since the electric company did that for me free of charge. It isn't hard to do, though; for every room you simply multiply the wall, window, door, floor, and ceiling areas times a certain heating load factor. Adding up the totals for each room gives the size of the furnace needed for the whole house. The heating load factor depends upon which area of the country you are building in. It can be obtained from a heating contractor or the power company.

212

The major cost of the forced-air central heating system is the ductwork, both for materials and labor. A furnace costs several hundred dollars; the pipes cost another several hundred. The labor you donate is worth more than several hundred dollars.

With forced-air heating, adding air-conditioning is a simple matter, once the ductwork is in place. You will have to buy a condenser unit, a compressor, and two lengths of special hose. The condenser is installed inside the furnace. The compressor unit houses the compressor and the evaporator coil, and is installed outside the home. The special hoses, prepressurized with *Freon gas*, connect the condenser with the compressor. These air-conditioning assemblies are available in various capacities to meet your needs.

INSULATION

The furnace is only half the heating system. The other half is so important I'm giving it a special section. *Insulation* keeps the heated air in the house after the furnace warms it. The trouble is that it really only slows down the heat loss as it's commonly used; it doesn't begin to stop it completely.

The cheapest way to insulate your house is to nail on some insulation board, which comes in 4- by 4-foot or 4- by 8-foot sheets, between the walls or roof sheathing and the building paper. Normally, such insulation comes in 1- or 2-inch thicknesses. At that thickness, it's genuinely cheap, inferior stuff, used mainly in apartment buildings.

In the long run, heating costs will be lower if you use another kind, *batting* or *blanket* insulation. This is like a quilt, $2\frac{1}{2}-6$ inches thick, depending on your finances. (Six inches of thickness is about twice the price of 2½ inches.) It goes between the wall studs, roof rafters, or floor/ceiling joists.

Until the energy shortage was a popular concern, 6-inch insulation was used only in roofs or floors, with 3½-inch insulation being the most common wall filler. Both have one side coated with aluminum foil, directed to the inside of the house to stop radiant heat losses. The foil side is stapled to the studwork.

In a house with exposed ceilings that are also the roof decking, as in my house, the quilt-type of insulation isn't applicable, since there isn't any finished ceiling board to cover them on the inside. Insulation board is the best alternative with cedar shakes, which are insulators themselves.

Actually, once the walls, ceiling, roof, and floor are insulated, there's much less heat lost through them. Windows and sliding-glass doors are the basic heat-wasting culprits in an otherwise properly insulated house.

CONSERVING ENERGY

It should be simple to build a house that uses little energy. Most household energy is used for heating or cooling the air inside the house. In both cases the goal is merely to keep the air at a comfortable temperature, regardless of the temperature outdoors.

If the air outside is too cold, the inside air is heated; if the air outside is too warm, the inside air is cooled. The problem is that heat moves around; if it's hotter indoors than outdoors, the heat will go through the walls until the outdoor and indoor temperatures are the same. If it's hotter outside than in, the heat will move the opposite way. Mother Nature doesn't want us to have our own special temperature spots; wherever we make one place have a different temperature than another place, the heat moves one way or the other to equalize the temperature in both places.

We have two ways to try to get what we want, however. The first is to make barriers in walls, floors, and ceilings that heat won't go through. Since all materials resist heat flow to some extent, most houses already pay at least token homage to this approach, although few are near the possible maximum efficiency. The second way is to heat or cool the air inside the house as fast as the heat slips through the walls to the outdoors.

Traditionally, houses have been built with large furnaces and thin walls, favoring this method because it's cheaper for the builder. But now that energy is recognized as being scarce and costly, a logical first step toward energy conservation in housing is to concentrate on making the whole house impervious to heat loss as much as possible.

With the proper construction and design, the furnace could be replaced with a small heater that need only operate part-time. A well-placed skylight could supply the necessary heat from sunlight on a sunny day. The stove or heater could be powered by wind or water-driven generators. Once the house is built to hold in every drop of heat, the problem of heating can be solved without much energy.

Building a house that holds heat isn't difficult; all the necessary products are commercially available right now. It's just expensive, perhaps $500 to $1500 extra initial cost for an average three-bedroom, 1500 square foot house.

To make such a house, you need to take into account the few rules about how heat works. First, you can't completely stop heat from going through the house, although you can slow it to a near halt. The house will always need a heat source of some kind, but the goal is to construct the house so the furnace needed is quite small. Heat goes through different materials at different speeds—it goes through metal quickly, glass less so, and wood or fiberglass quite slowly. Thus, the walls should be of a dense material with a high insulation factor—doors and windows should be made of special insulating material.

Second, the speed at which heat goes through any given material depends on the temperature difference between one side of the material and the other. If a box is made of inch-thick wood and the temperature inside the box is 10° lower than outside, the heat will flow into the box faster than if the temperature difference were only 5°. Any hot spots where heat accumulates should be either well insulated or eliminated by design; high ceilings, for instance, being always hotter than the room below, accelerate the heat lost through the roof unless added insulation is installed.

Third, hot air dissipates into cold air; whenever you open a door or window to the outdoors, the hot air rushes out. Even when the doors and windows are seldom opened, the inside air seeps out through cracks around doors and windows and through the walls themselves. Weather-stripping all openings and building all the walls, floors, and ceilings with care are the best protection against such losses, although they can't be entirely eliminated.

The best kind of house for conserving energy would be a simple cube, built with thick walls made of highly insulated material with a low roof, no doors and no windows—a giant oven.

A realistic approximation to this ideal is a common wood-frame house with floors, walls, and ceilings insulated with foil-backed 5- or 6-inch fiberglass insulation, triple-pane insulating windows, 2-inch insulated wood exterior doors (with storm-door covering), and thorough weather-stripping at all openings. Cheaper ways exist for making insulation like this, but none are as durable, easily available, or easy to install.

The average house, with 15% window area and two outside doors, could save 75% of the heat going through the walls, floor, and windows with proper insulation, assuming that no one starts opening and closing the doors like a madman. Such savings for an average house result in a need for only 10 Btu per square foot. Btu is a unit of heat, the *British thermal unit*; 1 Btu is equal to 3.41 watt-hours of energy. A single room-size heater could heat the entire house to 70° when the temperature outside is only 10°.

To illustrate what a dramatic effect the outdoor temperature has on the indoor temperature, assume you're heating a 20- by 50-foot living room with an 8-foot ceiling. A furnace would have to provide 6000 Btu (1800 watts) to heat the room. Eighteen 100-watt light bulbs would heat the room to 70° with the outside temperature at 10°. A 10-foot ceiling adds 25% more air volume and 25% more wall-surface area—25% more heat lost through the walls. A 12-foot ceiling adds 50% more heat loss; a 14-foot ceiling, 75%; a 16-foot ceiling, 100%.

In a more temperate climate such as the Pacific Northwest, where the average daily temperature is 40−50°, making the indoor/outdoor difference only one-third to one-half the 60° figure, the needed heat would be one-third to one-half the 6000 Btu: 3000 Btu corresponds to 900 watts. Burning three or four 250-watt light bulbs would keep the room warm.

A busy household might even generate enough heat as a byproduct of its normal activities; the hot water heater, oven, refrigerator, electric lights, clothes dryer, and other appliances give off plenty of heat as they operate.

Commercially available insulated windows and doors are costly; at least twice the regular price. Savings can be had by ordering the insulated glass from a glass company and making the window frames separately, bringing the frames to the glass company to have the glass installed.

Moving In

Ten months after applying for the building permit, eight months since my plans to finish had been shattered, the house was done. Ten thousand nails, tons of wood, and a cast of thousands of other parts were hiding in the silent, hollow box.

FINISHING

The basic house was done, all major parts complete. It seemed more than ever that we should be able to move in; three months had passed since the framing was completed; six since my first plans predicted I'd be done. I thought back on all the microscopic hours of work I'd done to whittle this enormous structure from raw materials. Only tiny details lay ahead, insignificant chores requiring only a few days' work. Someone else would do the inside walls. Cabinets, plumbing fixtures, window frames, molding and trim, interior doors, and flooring all together only made a tiny bit of work compared to that already done. I'd be done in another week or two. This goal in mind, my enthusiasm climbed; I'd have a last-burst effort and finish quickly.

Bathtub

First on the agenda was the bathtub. Long before, I'd spoken to a plumber who, on hearing I planned to do it all myself, had made a single warning: install the bathtub before the walls are complete; it won't fit through doorways and, if done later, walls will have to be torn out to get it in. I'd delayed it anyway, until now. I borrowed a pickup truck and ventured into the industrial downtown area to get the tub. I brought it to the house and a husky friend helped me wrestle the bulky, heavy thing down to the house. We set it down. There it stood, the first finished house part to arrive, gleaming in all its white enameled fiberglass splendor. I looked over it like a proud papa. But wait, it wasn't the one I ordered—the drain was on the wrong end.

The wholesale place refused to take it back. Luckily, the manufacturer was located nearby, and agreed to exchange it for the type I needed, with the drain in the right place; but I had to bring it back myself. Back to the pickup truck, we carried the awkward thing back *up* the hill—a day's work all by itself—and set off for the factory. We exchanged tubs and returned home, wrestling the new tub down the hill, this time much less carefully.

The troubles weren't over; it wouldn't fit through the doorway to its upstairs home. The inside walls were, luckily, still studs; I ripped several studs out of an upstairs wall, separating the loft area from the living room's high ceiling. We hoisted the heavy monster with some ropes and pulleys (from the defunct system outside). Balancing on a thin beam that supported the living room ceiling, I tugged and pulled until it was up there with me. There I was, up on this beam, the bathtub perched like an elephant beside me. I expected the beam to collapse at any moment under this weight, but nothing so dramatic happened. We tugged the big white creature through the ruptured wall and into the bathroom; a two-hour job had taken three days.

Interior and Exterior Walls

I called the drywall contractor; he would be out in two days, he said. Just enough time to install the insulation blankets in the walls before he came, I thought. But no, the night before he was to start, midnight caught Diane and I frantically stapling insulation into place, violently coughing as the fiberglass particles filtered through the air into our lungs, enraged at my poor planning. We worked all night; the workers were to arrive at dawn. But after 10 a.m., I began to wonder where they were, how they could be detained so long. Diane returned home; I started the day's chores.

The workers didn't come that day, nor the next. And though I got some sleep waiting for them, their lack of punctuality further upset me by destroying my schedule. They finally came in several days, but took another several weeks to finish what could and should have been a two-day job.

I exploded at the boss. Yes, I had made mistakes in framing. Yes, the house was hard to get to, and yes, everything about the house was harder than average. But I never misled him on these points; I had trouble getting anyone to take the job and he had willingly accepted the contract. He said he was losing money, that he had bid too low. But my loan was already made; I couldn't up the price. Why did his professional outfit have to spite me by taking so long? Emotions, I guess. He didn't answer—he was angry too. I was helpless.

While they dawdled, I did other things outside. I finished the sewer line construction. It was spring; the soil was soft for digging, the weather was warm. I stapled insulation under the house floor to protect it and the plumbing pipes from cold weather. I rebuilt the pathway from the road to the house and reinforced its retaining walls with larger boards, skidding with the wheelbarrow to the entertainment of passersby.

The siding, which Diane had practiced hammering on as she applied the whole front side (Fig. 13-1), was still left to do. I finished it in several days of hair-raising gymnastics atop a 35-foot ladder, fondly thinking of her, elf-like in the balloon parka and funky blue hat—*her* carpenter's uniform—and glad that I could end this risky high-altitude carpentry.

I settled into a confused emotional state while working; everything was moving like molasses. A part of me didn't want to finish at this point. House construction was my occupation; I had a job to come to every day, despite the boredom. I didn't have a boss. I could come and go as it suited me. No one could tell me what to do. I was succeeding at this housebuilding business. I had learned the hard way, and it seemed as if I

Fig. 13-1. Applying siding (cedar shakes, in this case) over the building paper that covers the exterior side of the framing walls.

ought to be able to rest on my laurels. When the house was finished, I'd have to leave the dream world of my project and come down to earth. I had no other projects in mind; I'd be simply back at a job. I didn't know what to do with my life. Despite all its horrors, housebuilding was something I had, by now, come to feel comfortable doing.

But I knew I didn't want to make it my profession; it remained a project for me. By not finishing the house, though, I could stave off decision.

Yet the greater part of me was desperate to finish; abandoned, half-done houses loomed through my night dreams and daydreams. And yet the finish was at hand. This tiny pile of details blocked my path, frustrating, tantalizing, and upsetting.

Diane began painting the inside wallboard, now completed, after spending several days shopping for the proper colors and prices. I retrieved the prebuilt cabinets and countertops after still more days of shopping for the best buy. People, friends and otherwise, filtered through the project, envious and awed, sympathetic and sometimes helpful. Buying took as long as building it now seemed.

ODDS AND ENDS

Thrift was now imperative. Even after all the trouble obtaining the loan, we fell about $1000 short; we had to get the money we needed from Diane's income. All spending was cut to the bone; we became vegetarians, quit going out for entertainment (one movie a month was our allowance), gave up soft drinks, hard drinks, and other expensive habits, and saved every penny for the house.

Cabinets and Countertops

The work dragged on. A week was spent on cabinets, leveling, assembling, installing (from their enclosed directions), and modifying them to fit the built-in oven, stove, and sinks. Plumbing fixtures devoured more time yet—ordering, transporting, and unpacking them was a job in itself. Connecting the supply and drain pipes, correcting mistakes, and running to the plumbing stores each day—it seemed as if I'd never finish.

Kitchen cabinets are available in all price ranges and styles, prefinished. They save time and, if you're uncertain of your cabinetmaking ability, anguish. Premade countertops are the best answer to your lack of ability, too. Simply buy the cabinets, level them, nail them in place, and glue on the countertops. Any cabinet company will calculate the cabinet sizes and amounts you need; all you need to do is follow a plan and put everything in place.

Moldings

Fitting wood moldings for the doors and windows, inside and out, required more precision than my hands and saw could provide; I spent time finding a table saw to rent for these precision cuts. Installing inside doors between the rooms, and painting, cutting, and installing moldings took several weeks. The wall paneling in the living room required that each of the 400 one-by-six boards be cut to fit exactly, since the ends weren't factory-matched. I can't describe the tedium, to say nothing of scaling the 21-foot living room walls on my rickety ladder to place these finished pieces—three weeks for the paneling. These little finishing touches were driving me crazy.

Doors and Deck

The inside doors, though they came prebuilt, took a while to fit in place so that they opened smoothly. Hanging all the light fixtures took a couple of days, including time spent running back and forth to get forgotten parts. Even then the floor required another layer of lumber—underlayment, as it's called—to give additional strength and silence, and a uniform, smooth backing for the carpeting.

The deck required a deck railing. The mailbox was installed. I called the water bureau to connect the water from the street main. The sewer line was connected to the sewer by another contractor and the city sewer people. Mirrors had to be hung. Electric outlets and switch covers were installed. Doorknobs, towel bars, and toilet-paper holders had to be put in—another few day's work.

Gutters and downspouts took almost a week of buying, hauling, painting, assembling, and installing. Outdoor lighting included digging ditches for the buried wire; more buying, installing, and wiring consumed another several days.

If you're working on a two-story house, installing the stairway as soon as the second level is done will save an unbelievable amount of time and energy over the course of construction; I wasted much time and fatigue running up and down ladders.

A design detail to note is that joists are easily extended through exterior walls to make outside decks (Fig. 13-2); most of the labor goes into placing the joist, so that making it a little longer takes no more effort and costs little more. A deck gained in this way would be almost free.

Fig. 13-2. A deck can be added easily and inexpensively. Increase the length of the floor joists to extend through the exterior wall. Then cover the exposed portion of the joists with a finishing material suitable for outdoor conditions.

Floors

I laid some linoleum myself, but had an expert do the larger patches where seams had to be hidden. Then carpet installers came and finished the job, turning the dusty carpenter's workhouse into a residence.

Carpet and linoleum laying is an art. Except for small pieces where no seams have to be hidden, doing it yourself is a gamble on this, the most visible part of the house. It would be tragic to build everything yourself in a reliable way but have the part that shows come out ugly.

Tools

Organizing tools and materials can save a tremendous amount of time; I never did, though, and spent perhaps an hour a day looking for mislaid tools I needed.

225

Carpentry tools you need to buy are few. The absolutely necessary ones are: a claw hammer, chalkline and reel, framing square, carpenter's level, hacksaw, calking gun, plumb bob, pencil, power saw, hand crosscut saw, carpenter's utility pouch, and two sawhorses.

A table saw is needed for finishing work. A half-inch drill is needed for bolted connections, plumbing, and wiring; a quarter-inch drill is handy, too. These last three can be rented, though, as can a staple gun and staple hammer used to apply building paper and insulation.

The hammer should be at least 16 ounces; 20 ounces or more make things go faster, though. The carpenter's level must be accurate and 2–3 feet long. The portable power saw should be accurate, light, and easy to operate, since you'll be using it all day long.

CONCLUSION

Aside from the frigid roof experience and the beginning toil in the dirt, building the house was rewarding in several basic ways. Shaping raw materials into a functional wall, floor, or roof was gleefully exciting. Making saw cuts just right, driving 2½-inch nails in three hammer blows, having surfaces come out level without worrying about it, and getting things done quickly and easily were the simple joys I found.

I whistled and worked, watching my hands perform increasingly skilled tasks, and watched the house grow through my labors.

Sometimes getting lost in the mechanical action of hammering was a blissful trance-like thing; going faster and faster, everything clicking into place, I savored the experience as solid products flowed so easily from my own hands. Working at these crafts, where every motion improved my skill, was the most rewarding part of housebuilding.

But, unfortunately, single-person housebuilding required that I could only seldom fully live this more involving role. Usually, I had to be two people—*workman* and *contractor*. I was the workman who comes in on Monday morning, with hammer, nails, or whatever, knows what to do to put things together, and does so. And I was the supervisor who calls the lumberyard to order the supplies for the workman coming in Monday (myself).

I put the house together during the day, and oversaw the construction and material-buying details at night. The trouble was that the two roles couldn't be separated so easily; I had to purchase things during working hours. I had to make sure the workman was doing things according to plan. I got little construction work done on the house for days at a time, while I was out deciding what to buy or where to buy it, waiting in line at city hall to get a permit, driving across town to see if some piece of second-hand material could be of use, hunting from store to store for a rare part I had to have, fixing my ever-broken pickup truck, sitting in dejection—often sopping wet from rain—in the local coffee shop, visiting or calling the engineer to ask how this part fits that, hunting for a house in the same phase of construction so I could ask a workman how this part goes, waiting for materials to arrive if I made a mistake in ordering too late, or talking to someone about the weather. I was constantly forgetting something at the store and having to go back for it.

Unexpected things kept happening that the workman part of me turned over to the contractor. I made it a practice, for instance, to order less material than I needed, if there was a doubt, so that I wouldn't be stuck with anything I couldn't use. (Although most lumberyards will allow you to return most unused materials.) When I started building walls, I soon ran out of two-by-fours, and had to stop work and go to the lumberyard, wasting half a day getting what I needed.

Going to a store may not sound like much, but it was a problem all by itself. Walking into the store on any weekday morning, I'd find a dozen people waiting to be helped. A normal order might be several pounds of nails, a few sheets of plywood, and three 10-foot-long 3- by 12-inch boards. The clerks had to retrieve the order from the yard in back, which took a while. Everyone's purchases went on his individual account, so all was written down in detail, taking more time. It wasn't uncommon to spend an hour waiting and another 30 minutes getting my own order filled. By the time I drove to the store, waited for the order, and returned to the house, the spirit of the day was often broken.

After working alone for several months, the contractor got a little dejected, and had trouble setting strict, regular working hours. Starting my errands at 7:00 a.m., I would return to the house around 11:00 a.m. After a while, working alone began to seem like a hopeless battle and the working pace went slower and slower as the interruptions multiplied; sometimes several trips to the store were needed each day, so I'd sit in a coffee shop to end the workday, nothing being accomplished.

In order to avoid some errand-running, I finally developed a feeling for the way to work without a needed tool. At first, I tried to use the proper tool for everything: I wanted to get a *ditch-digging* machine to dig the sewer ditches, a *power auger* for drilling the foundation holes, a special concrete worker's *level* to make the foundation wall exactly level, a special saw to cut the angles for the rafters, and an *air nailer* to attach the flooring.

I spent two weeks searching for the ditch digger and the auger, and wound up using a shovel; I looked for a special level for several days, until I saw a concrete worker using an ordinary level; and I cut the rafters with my own saw.

As time wore on, I found the only way to get things done without wasting all my time on errands was to solve problems as best I could on the spot, without resorting to whatever imaginary "perfect" apparatus might exist to make it easier.

Building your own house can save a lot of money, even if it first appears otherwise. Taking a year off from conventional employment, even though you save half the cost of a new house, may cost as much in lost income, losing $10,000 income so you can save $10,000 on a $20,000 house gains nothing, it appears, until you see how the mortgage costs differ.

A $20,000 mortgage at 7% interest, for example, will cost more than $40,000 over a 25-year repayment time. A $10,000 mortgage at the same interest rate and time period will double too, to $20,000 total cost. The difference in cost, then, between a $10,000 house and a $20,000 house is not $10,000, but *$20,000*!

Since a mortgage often doubles the price of a house in the long run, saving half the purchase cost (half the mortgage) means saving the whole purchase cost in the long run!

Having built the house yourself, you'll know how to repair it, too. No more plumbers' or electricians' bills to pay. You'll be an all-around handyman, able to help others build houses if that's your wish.

Index

Index

A

Attic   133

B

Backhoe   83
Baseboard heater   210
Bathtub   219
Batting   213
Beams   73
    hanger   122
Board foot   57, 60
Bridges   142
Building
    codes   37
    inspector   22
    paper   132
    permit   47
    restrictions   36

C

Cabinets   223
Carpets   225
Ceilings
    joists   167
    materials   166
Clearing ground   51
Competitive bid   53
Concrete   70, 75
    characteristics   85
    footings   73, 75, 80
    pouring   84
    pump   84
    slabs   79
    utilities   77
    walls   75
Contractors   25, 87
Corner posts   107
Costs   28
Crawl space   73
Cripples   107
Cross bridging   119

D

Decorative framing   99
Doors
    cylinder lock   161
    hollow core   161
    installation   160
    solid core   161
Drawing fee   27
Ductwork   209

E

Exposed-beam construction   140

F

Financing   42
Fireplace   147
Floors
    blocking   99
    joists   101

subfloor	99, 110
Footings	73, 80
Foundation	38
Framing	90, 97
decorative	99
methods	110
skeleton	99
wall	104
French drain	50, 63, 128
Frost line	80

G

Glazing	156
Grade, lumber	61
Green lumber	62

H

Hanger, joist and beam	122
Heating load factor	211
Hollow core door	161
House construction	
advantages	12
complexity	13
satisfaction	11
time requirement	16, 24

I

Insulation	163, 213
batting	213
theory	214
weather-stripping	215
Interior	96

J

Jacks, wall	91
Joist	133, 166
hanger	122

K

Kern, Ken	29
Kiln-dried lumber	62

L

Linoleum	225
Loans	39, 44
Lock	161
Lot (def.)	35

Lumber

grade	61
green	62
kiln-dryed	62
new	60
used	61

M

Map	35
Minimum lot size	36
Moldings	223
Mortage	228

N

Nailing strips	120
No-hub pipe	170
Normal foundation	38

P

Paper, building	132
Paper, roofing	132
Partitions	107
Patches	73
Permit	22, 37
plumbing	129
Piling foundation	38
Pipe	181
Plans	
fee	27
limitations	19
modifications	20
Plumbing	
design	178
finished	175
fittings	181
hints	190
permit	129
pipe	181
rough	175
traps	183
vents	183
Post	
anchor	121
-and-beam construction	141
Posthole digger	83
Prefabricated houses	
costs	27
inadequacies	27
time requirements	28
P-trap	172
Puddling	72

R

Rafters	132
Ridge beam	137
Roof	
attic space	133
bridges	142
building paper	132
ceiling joists	133
exposed beams	140
outer	129
paper	132
post-and-beam	141
rafters	132
ridge beam	137
skeleton	129
shingles	131
trusses	133
types	133
vents	132
Rough plumbing	171

S

San-tee fitting	181
Septic tank	34
Sewer	34
Shakes	125
Sheathing. wall	117
Shingles	129
Siding	164
Site	
codes	37
costs	38
financing	39
finding	38
location	38
pricing	33
requirements	34
restrictions	36
shopping	31
zoning	32. 35
Skeleton framing	99
Slabs	79
Soldering	174
Solid core door	161
Square	126
Stairway	224

T

Takeoff	52
Tax records	38
Toenailing	120
Tools	69. 89. 225
Traps	183
Trusses	133

U

Used materials	57

V

Vents	132. 183
Vibrator	86

W

Walls	
framing	104
installation	163. 165
jacks	91. 117
materials	162
opening	106
posts	107
sheathing	117
siding	110. 164
Windows	
casing	157
frame	156
glazing	156
installation	158
prebuilt	155
styles	158
Wiring	
color code	202
fundamentals	199
hints	207
planning	203
utility box	201

Y

Yards. concrete	84